Botanical Diversity in the Tropical Rain Forest of Guyana

The Tropenbos-Guyana Programme operates within the framework of the international programme of the Tropenbos foundation and is executed under the responsability of Utrecht University. The multi-disciplinary Tropenbos-Guyana Programme contributes to conservation and wise utilization of forest resources in Guyana by conducting strategic and applied research and upgrading Guyanese capabilities in the field of forest-related sciences.
The Tropenbos-Guyana Series publishes results of research projects carried out in the framework of the Tropenbos-Guyana Programme.

R.C. Ek
Botanical diversity in the tropical rain forest of Guyana
Tropenbos-Guyana Series 4
Tropenbos-Guyana Programme - Georgetown, Guyana
ISBN: 90-393-1773-9
Keywords: Botanical diversity, species richness and abundance, logging, logging damage.

Printed by Elinkwijk bv
Cover Photos and computer image by Renske Ek.
Frontpage and background: Lianas in Greenheart forest, Pibiri, with e.g. *Connarus perrotteti* var. *rufus*, *Moutabea guianensis* & *Lonchocarpus negrensis*.
Backpage: Administration of logged Greenheart tree, Pibiri.
Computer image: Presentation of exploited one-ha plot, Waraputa.
Invitation: Young palm, Mixed rain forest, Saül, French Guiana.
Lay-out Bart Landman.

Botanical diversity in the Tropical Rain Forest of Guyana

Botanische diversiteit in het tropisch regenwoud van Guyana

(Met een samenvatting in het Nederlands)

Proefschrift

Ter verkrijging van de graad van doctor
aan de Universiteit Utrecht,
op gezag van de Rector Magnificus,
Prof. Dr. H.O. Voorma,
ingevolge het besluit van het College van Decanen
in het openbaar te verdedigen op
donderdag 4 December 1997
des middags te 12.45 uur

door

Renske Cornelia Ek

geboren op 24 December 1960 te Dordrecht

Promotoren: Prof. Dr. M.J.A. Werger
Hoogleraar in de botanische oecologie en vegetatiekunde, Universiteit Utrecht
Prof. Dr. P. Hogeweg
Hoogleraar in de theoretische biologie, Universiteit Utrecht

TROPENBOS

Universiteit Utrecht

The investigations reported in this thesis were carried out at the Mabura Hill fieldwork site of the Tropenbos Guyana Programme, 12E Garnett Street, Campbelville, Georgetown, Guyana and at the Departments of Botanical Ecology and Evolutionary Biology and Theoretical Biology, Utrecht University, P.O. Box 800.84, 3508 TB Utrecht, The Netherlands.
e-mail: r.c.ek@pobox.ruu.nl

Funding was provided by the Tropenbos foundation and the European Commission.

All she wanted was her three wishes granted. Nothing fancy: to be the new Tarzan of the 20th century, to contribute to wise management of tropical forest, to save the world, the usual stuff

What she got was something completely different:

'Jungle surrounded her. It wasn't nice, interesting, open jungle, such as leopard-skin-clad heroines might swing through, but serious, real jungle, jungle that towered up like solid slabs of greenness, thorned and barbed, jungle in which every representative of the vegetable kingdom had really rolled up its bark and got down to the strenuous business of outgrowing all competitors. The soil was hardly soil at all, but dead plants on their way to composthood; water dripped from leaf to leaf, insects whined in the humid, spore-laden air, and there could be this terrible breathless silence made by the motors of photosynthesis running flat out. Any yodeling heroine who tried to swing through that lot might just as well take her chances with a bean-slicer.'

After: ~~Faust~~ Eric. A discworld novel, by T. Pratchett, 1990

Contents

CHAPTER I

General introduction

'A proper understanding of the most relevant aspects of biodiversity in tropical rain forests, as well as the ability to predict the impacts of human intervention on biodiversity, are imperative to conservation efforts of the biological richness of these forests' (Tropenbos 1996)

This statement is found in the policy paper concerning priorities in biodiversity research by the Dutch Tropenbos Foundation, and encompasses the core of the study on 'Botanical Diversity in the tropical rain forest of Guyana', presented in this thesis. It introduces a large part of the terminologies used in relation with biodiversity in tropical rain forests:

Tropical rain forests: Tropical rain forests are the evergreen forests of the humid lowlands and hills occurring roughly between the tropics of Cancer and Capricorn, in areas where the annual amount of rain exceeds 1500 - 2000 mm per year (Huston 1994). These forests cover only 7 % of the earth's surface and are estimated to contain more than half of all the species of living organisms on earth (Wilson 1992). Tropical rain forests were initially regarded as stable and unchanging environments where the tropical plant community was at equilibrium. Lately, however, they are considered to be much more dynamic. They are subject to a variety of disturbances and environmental fluctuations, both with small-scale, short-term effects and with large-scale, long-term effects (Huston 1994).

Impacts of human intervention: At this moment, considerable parts of the world's tropical forests are altered by commercial logging activities. Along with the cutting of tropical rain forests for agricultural lands, these logging activities are considered to be the main cause of the rapid reduction of tropical forests (Myers 1988, 1989; Werger 1992; Bruenig 1996).With the disappearance of these forests, animal and plant populations may disappear, in extreme cases leading to extinction. The attention that this issue received resulted in the introduction of a new term: biodiversity.

Biodiversity: The term 'biodiversity' is used in several disciplines and many different definitions exist. In general, these definitions can be grouped under three headings: biodiversity as a concept, as a measurable entity and as a social or political construct (Gaston 1996). As a concept, biodiversity encompasses the total variety within the living world, ranging from variation at the molecular and genetic level, through variation at the species level, to variation in functionally related groups of species and their habitats at the landscape level. As a measurable entity, the number of species is the most widely applied measure, implicating that variation at the species level and at the landscape level are the most widely evaluated levels of biodiversity. This explains why, in many discussions concerning biodiversity, species richness is often considered a synonym of biodiversity. Species diversity is described by two parameters: *species richness*, which is number of species per unit area and *species abundance*, which is number of individuals of a species per unit area. Lastly, as a social or political construct, the term biodiversity focusses on the effects of changes in land use, associated with changes in species numbers, and in particular on the impacts that these changes in biodiversity will have on the quality of the human environment, both physically and spiritually. The global social and political concern over the decrease in biodiversity, degradation of forests and social inequality initiated the Convention on Biological Diversity of the United Nations Conference on Environment and Development (UNCED) held in Rio de Janeiro in 1992 (Johnson 1993). That conference resulted in a treaty in which a series of recommendations was formulated to stop the destruction of species, habitats, and ecosystems, called 'Agenda 21'. The overall objectives of the treaty were 'the conservation of biological diversity, the sustainable use of its components and the fair and equitable sharing of the benefits arising out of the use of genetic resources' (EEI 1997). During the preparation phase of the treaty, and afterwards, many new policy initiatives concerning tropical rain forest biodiversity were generated. In relation to forest management, e.g., the International Tropical Timber Organization (ITTO) and the Centre for International Forestry Research (CIFOR) have initiated activities to incorporate the conservation of biodiversity as an additional objective in the management of timber production forest.

Biological richness: Biological richness is often used as a synonym for biodiversity. In general, biological richness in natural habitats is high in tropical lowland areas and decreases with increasing latitude and altitude (*e.g.*, Gentry 1990, 1993). This implicates that lowland regions around the equator are biologically rich.

However, this is not by definition true, not even for all tropical lowland forests. Diversity and richness vary with geographic region, abiotic and biotic environment, and geological, evolutionary and disturbance history (Fig. 1.1). As a result biological richness changes in space and time.

Conservation: As the term biodiversity is linked with a global concern on the disappearance of species and ecosystems, it is strongly connected with the issue of nature conservation. Biodiversity is viewed as a value, either out of its own rights, or as an economical 'option value'. The option value represents the potential value of biodiversity, ensuring possible future uses, thus actually possible future threats, to biodiversity.

Relevant aspects of biodiversity: The option value discussed above does not, however, touch on the complex topic of how biological diversity or richness affects the functioning of tropical forests. What is the significance of having a particular number of species participating in ecosystem processes, such as cycling of organic matter and nutrients? Functional groups, *i.e.* groups of species that participate in a particular process, can differ in number of participating species. Many species are not per definition better than few species for carrying out, or maintaining, a particular ecosystem process. Ewel *et al.* (1991) and Lugo (1992), for instance, found that diverse natural successions and tree plantations of a single species were similar in their ability to maintain soil fertility. Furthermore, Brouwer (1996) found decreasing nutrient leaching with increasing vegetation cover after logging, even if this vegetation cover would be constituted merely by *Cecropia* spp. Although a few species may have a similar effect on the functioning of a specific ecosystem process than larger numbers of species, species may differ, nevertheless, in their habitat requirements, reproductive characteristics, growth rates, and mortality rates and causes. This then means that species composition has an effect on the range of biotic variability. Species composition may have a direct effect on many ecosystem processes, *e.g.* rates of herbivory (Waterman & McKey 1990), and rates of litter production and decomposition (Waterman *et al.* 1980; Van Schaik & Mirmanto 1985; Veneklaas 1991; Burghouts *et al.* 1994). Species diversity may add robustness in ecosystem processes (Waterman & McKey 1990; Ewel *et al.* 1991; Lawton & Brown 1993; Vitousek & Hooper 1993; Tilman & Downing 1994). In other words it buffers impacts and helps in adaptation to new situations. Recently, experimental results of Naeem *et al.* (1994, 1995) showed that uptake of carbon dioxide and plant productivity (produced biomass) declined with declining species richness. This seems to indicate that high species richness improves ecosystem functioning of tropical forests.

Trying to understand the most relevant aspects of biodiversity, and to be able to predict the impacts of human intervention on biodiversity: An attempt to understand the most relevant aspects of biodiversity in relation to logging, would firstly require a description of the characteristic species diversity of the forest. Secondly, an evaluation of the impacts of logging on this characteristic species diversity is needed. If possible, functional groups that show responses to logging that are highly correlated with logging damage or groups with a possible key-stone function should be identified. Lastly, it has to be analyzed if it is possible to predict the responses of the forest community to logging, or if the processes that will follow exploitation are largely governed by random factors. In this thesis, a study is presented of the effects of logging on the biodiversity of Greenheart dominated rain forest in Guyana, South America. The study is partly linked to the 'Logging Experiment' (Van der Hout 1996), a reduced-impact logging project designed to develop a sustainable forest management system for Guyana's Greenheart forest. In this study 'sustainable logging' indicates the sustained production of commercially interesting timber species, combined with maintenance of the characteristic botanical species diversity of the forest.

Tropenbos: The Dutch Tropenbos Foundation aims at contributing to the conservation and wise management of tropical rain forests worldwide by generating tools for policy makers and managers. The second main objective is to involve and strengthen local research institutions and capacity in relation to tropical rain forest (Tropenbos 1996, 1997). Under its supervision, research programmes are being executed in Cameroon, Colombia, Côte d'Ivoire, Indonesia, and Guyana. The programme in Guyana started in 1989, and is a joint research programme of the governments of Guyana and the Netherlands. The Tropenbos-Guyana programme aims at developing guidelines for sustainable forest exploitation and forest conservation. To study indigenous forest uses, with a focus on non-timber forest products, a new study area was opened in 1995 in the North-West district of Guyana. The main part of the programme, however, is concentrated in a logging concession in the central area of Guyana and focuses on the sustainable exploitation of timber. The concession is presently managed by Demerara Timbers Ltd. (DTL). A multi-disciplinary research team of Guyanese, Dutch, and English institutions studies the abiotic and biotic components of the forest ecosystem and the processes between the components. The results of the first phase of the research programme (1989-1993) and resulting recommendations for management are summarized by ter Steege *et al.* (1996).

Botanical Diversity

The processes that lead to present-day levels of (plant) species diversity are considered to be a combination of both historical (origin of diversity) and present-day ecological processes (maintenance of diversity) (Fig. 1.1). Geological and evolutionary history, climate and rainfall are considered important at a large spatial and temporal scale, while microclimatic or disturbance-based processes are the most frequently mentioned present-day ecological mechanisms explaining differences in diversity on smaller temporal and spatial scales (Huston 1994). It can also be argued that the term 'disturbance' is already applied to many different spatial and temporal scales, and almost all processes can be seen as processes of disturbance. In that case geological history (*e.g.*, continental drift, glacial periods) and evolutionary history (*e.g.*, speciation and extinction rates, and patterns of dispersal and immigration, level A in Fig. 1.1) can be considered on the level of mega-scale disturbances (Haffer 1974, 1982, 1997; Prance 1982; Gentry 1993; Colinveaux 1993). Furthermore, landslides, volcanic eruptions, earthquakes, hurricanes, fires, drought, and large scale river dynamics (Salo *et al.* 1986; Kalliola *et al.* 1991) can be placed on a large to intermediate spatial scale, but smaller temporal scale of disturbance (after Waide & Lugo 1992). Lastly, the collapse of aged or diseased trees (gap dynamics theory) can be seen as a disturbance on the smallest spatial and temporal scale (*e.g.*, Connell 1978; Brokaw 1985a, 1985b; Popma & Bongers 1988; Van der Meer & Bongers 1996a; Zagt 1997). The concept of disturbance as mentioned as level C in Figure 1 would be applicable to the last two levels of disturbance.

The different abiotic factors: soil fertility, hydrology, temperature, and light (level B in Fig. 1.1), are all considered critical variables in tropical forests which are directly correlated with the level of plant diversity (*e.g.*, Gentry 1988; Wright 1992; Huston 1994; Clinebell *et al.* 1995). Additionally, they may influence the impact or frequency of a disturbance. For instance, trees on shallow, wetter soils are expected to show higher turnover rates than trees that grow on more deep and less soggy soils (Kapos *et al.* 1990). Also, in areas with limited water availability, a fire that originates from lightning might have much larger spatial impact than if that same lightning had struck within a vegetation with much higher water retention capacities.

The process of gap dynamics and tree turnover rates is also an example of how biodiversity itself can influence the level of disturbance. The biotic microclimate and self-regulatory processes of plant diversity (*e.g.*, competition, reproduction, growth, and mortality) are indicated as level D in Figure 1.1, and given in more detail in Figure 1.2.

Botanical Diversity and Logging

Logging involves the selective extraction of trees. Logging will lead to the creation of gaps (*e.g.*, Connell & Lowman 1989; Ter Steege *et al.* 1996), and the mechanized transport of logs by skidders will lead to soil compaction, uprooting of the soil by over-run vegetation and damage to the remaining stand (Ter Steege *et al.1996;* Bruijnzeel 1990; Bruijnzeel & Critchley 1994). Especially the skidder activities will lead to impacts dissimilar to those of any natural disturbance. On skid trails the hydrological characteristics, such as water retention capacity and hydraulic conductivity of the soil, chance (Jetten 1994), and the absence of sufficient living roots has been shown to lead to leaching of nutrients (mainly cations) (Brouwer 1996). Furthermore, soil acidification and raised Aluminum levels may occur (Brouwer 1996).

Logging will have a direct impact on plant diversity, operating through changing growth and mortality rates of individual plants (Fig. 1.3). This chance may lead to

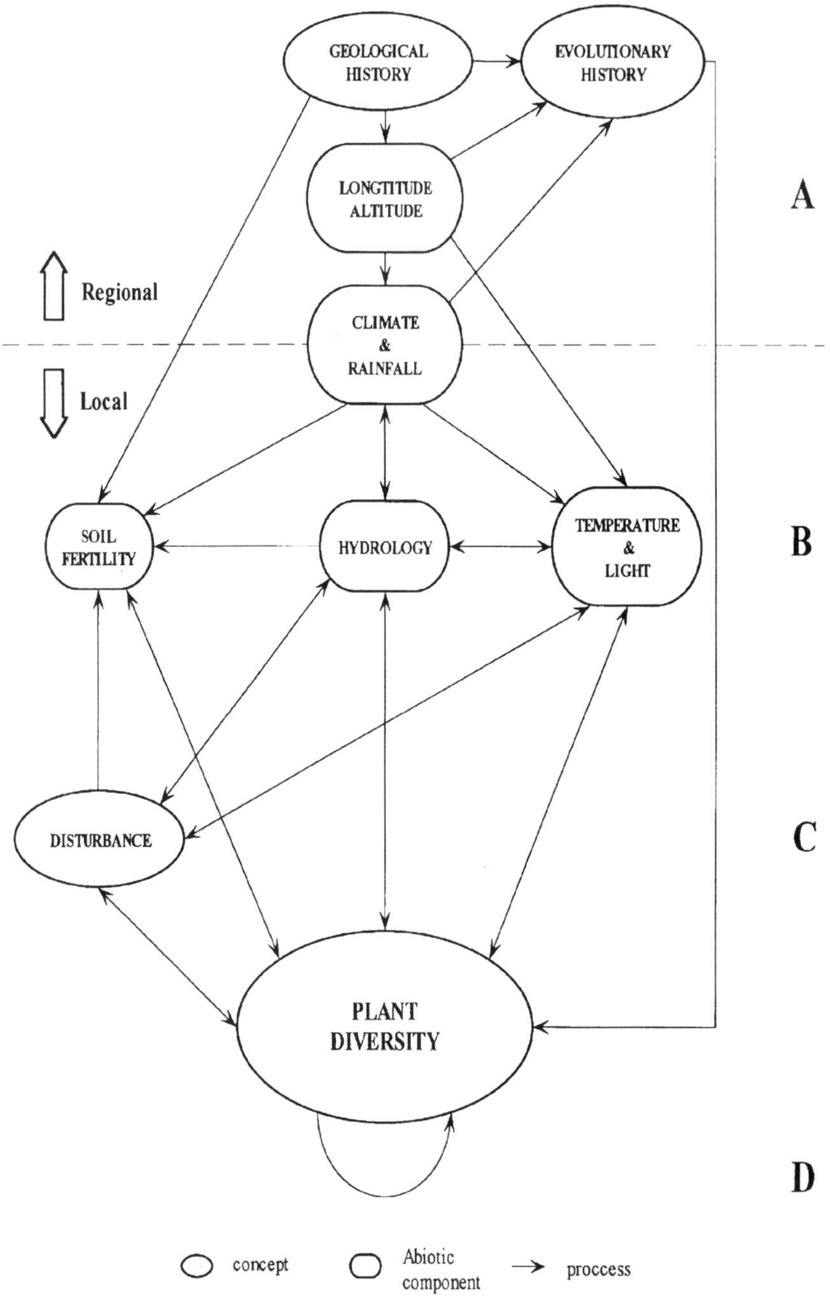

Figure 1.1
Simplified diagram of factors and interactions that influence the present-day levels of plant diversity.

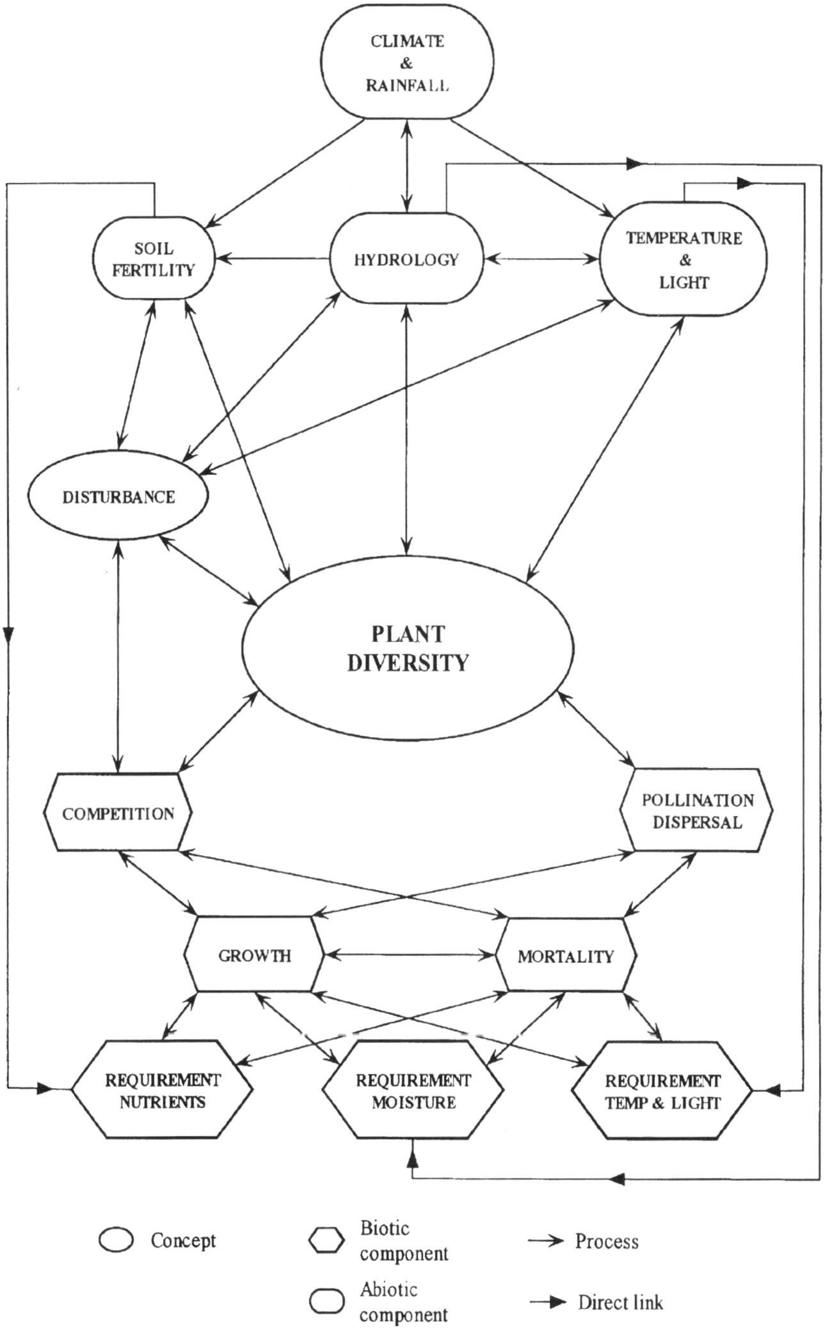

Figure 1.2
Simplified diagram of factors and interactions that influence the present-day levels of plant diversity without geological and evolutionary history, and with emphasis and the interactions between the biotic and abiotic environmental factors.

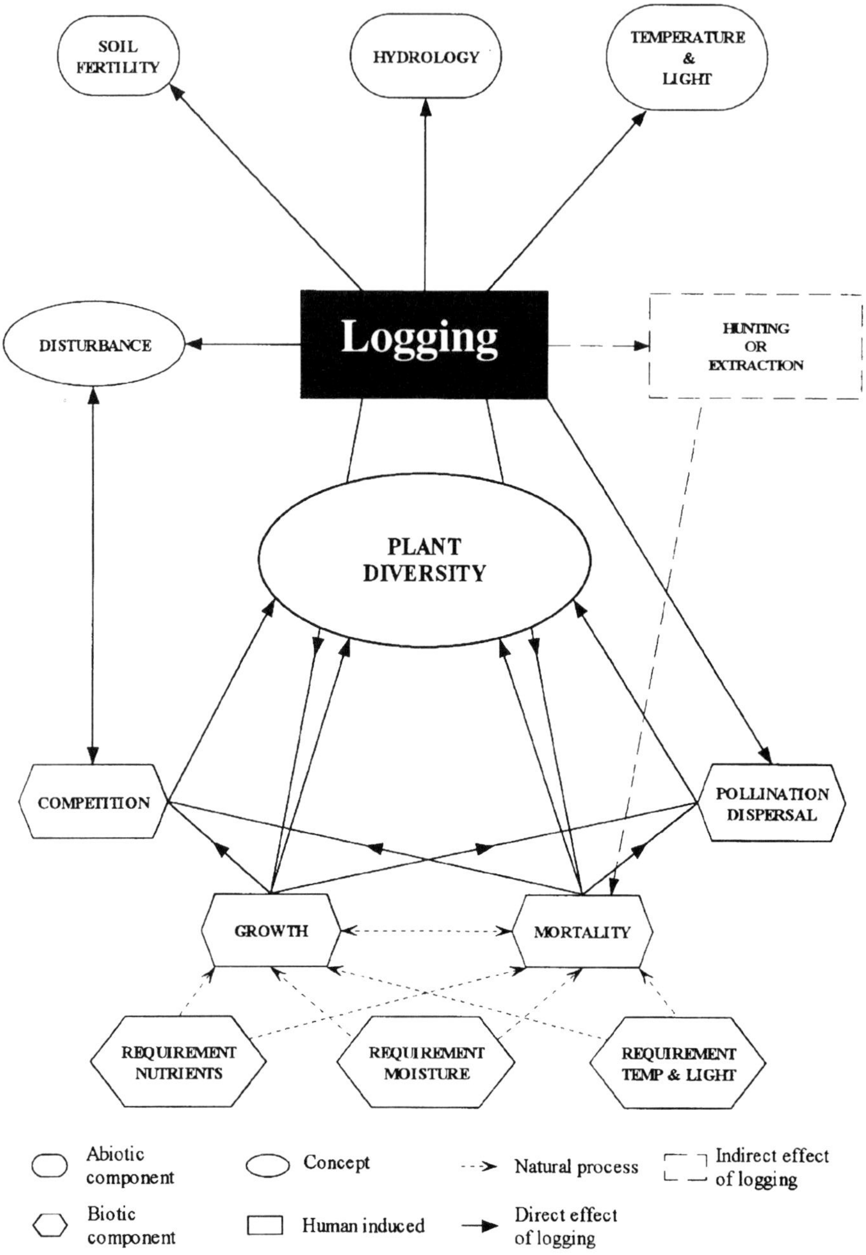

Figure 1.3
Simplified diagram of the direct, and possible indirect, influences of logging on levels of plant diversity.

consequences on a within-species level (*e.g.*, changing the genetic variability of the harvested tree species), through changes of population structure. It may also lead to changes in species composition of the forest. Compared to natural disturbances, which usually occur more frequently than logging (after Waide & Lugo 1992), logging may have an effect on a larger spatial scale. Logging may also have an indirect impact on the biodiversity (Fig. 1.3). For instance, it makes the forest more accessible for hunters or gatherers, which may lead to more 'harvested' species. Normally, herbivores such as agouti, peccary, and tapir, assist in the dispersal of plants and/or may prevent the dominance of particular plant species or groups (*e.g.*, Janzen 1970; Connell 1971). These herbivores in their turn are predated by *e.g.*, jaguars. When hunters decrease the populations of these herbivores or their predators, this may also have consequences on the forest structure and diversity (Terborgh 1992). Gatherers, for instance of living parrots or non-timber forest products, may also reduce local plant populations (Peres 1990; Thiollay 1989; Vásquez & Gentry 1989). Another indirect impact of logging may be the isolation of forest patches for specific pollinators and dispersers. Furthermore, through a direct influence on the abiotic components of the microclimate, the available levels of nutrients, moisture, temperature, and light may chance, changing the competition levels between species concerning those requirements. Smith & Huston (1989) showed, based on individual-based models, that no organism can be dominant over the entire range of conditions under which it can survive (from Huston 1994). If the abiotic conditions chance, other species will become dominant, and may alter the species composition.

Selective logging may lead to a series of changes which separately, or in combination, lead to changes in forest structure and diversity. Changes in microclimate, possible invasion of species from nearby secondary forest or forest edges, possible fragmentation of the forest, possible increased tree turnover rates after logging, and changes in strength of competitive interactions, may alter the composition and relative abundances of species occurring in Greenheart forest (Fig. 1.3).

Recovery of the forest after disturbance

Under natural circumstances, between 0.5 and 3.6 % of the trees (dbh > 10 cm) per year die off in tropical rain forests (Phillips *et al.* 1994). Natural gaps created by those tree falls rarely exceed more than 500 m^2 (Denslow 1996), and when large gaps occur this is mainly due to the fall of a large-crowned tree or the enlargement of smaller gaps (due to *e.g.*, the effects of lianas or wind turbulence (Hubbell & Foster 1986)). In contrast, gaps created through felling are usually much larger. A tree that dies from natural causes has a smaller or already decayed crown (Van der Meer & Bongers 1996a). The harvested tree, however, is mostly a nice full-grown specimen, with fully developed branches and full of leaves. Changes in the environment due to gaps are in general moderated by remnants of the previous community and by the surrounding forest (Denslow 1996; Brouwer 1996). Light, humidity, wind, and temperature are tempered by the adjacent tall forest, and remnants of the previous vegetation, such as soil organic matter, seed stocks, seedlings, shrubs, and saplings, ensure that soil processes and forest community composition remain modified to a minor degree in small gaps (Jordan 1985; Vitousek & Denslow 1986; Uhl *et al.* 1988; Van der Meer & Bongers 1996b; Brouwer 1996). This modification is larger in larger gaps.

Recent research in French Guiana showed that vegetation recovery of small and large gaps operated differently. Small gaps seemed to fill mainly by means of lateral ingrowth of surrounding trees and large gaps filled mainly through growth from the gap floor, both of colonizers and survivors from before the disturbance (Van der Meer & Bongers 1996b). The increased light availability in and around gaps is believed to be the most important factor controlling seedling establishment and performance (*e.g.*, Clark & Clark 1990; Welden *et al.* 1991; Zagt & Werger in press), thus determining to an important extent the vegetation composition after forest recovery (*e.g.*, Connell 198; Brokow 1985; Popma & Bongers 1988). The light conditions within gaps (Clark & Clark 1990) and in their immediate surroundings (Popma *et al.* 1988; Lieberman *et al.* 1989; Van der Meer & Bongers 1996b) may vary considerably as a result of variation in gap size, form, and orientation towards the prevailing course of the sun (Ter Steege 1993). Thus, gap size and form seem to determine the chance in light environment after a disturbance, and the moderating impact of the surrounding forest. Logging leads to larger gaps, and due to high logging intensities (traditional logging in Guyana up to 25 trees/ha, pers. obs), many of those large gaps are created at the same time. Furthermore, skidding activities influence and damage the surrounding forest, decreasing its moderating and buffering impact.

Logging leads to increased establishment and rapid growth of individuals within gaps (*e.g.*, Brokaw 1985b; Denslow 1995). The favoured individuals in and around the logging gaps can belong to species already present before the logging, as well as to invading species. Even if no new species arrive after logging, the densities of individuals may increase in large gaps, increasing the chance of higher species richness per unit area (after Denslow 1995). If the effects of logging are evaluated on spatial scales that are too small, it can be expected that at all levels of intensity or spatial extent of logging, logging may lead to an increased species richness due to this increased densities of individuals.

Whether or not the disturbed forest will be able to restore its former level of botanical diversity depends on the outcome of succession in the damaged forest patches. The amount of remnant surrounding forest, availability of seed vectors in this surrounding forest and their arrival time, the availability of suitable establishment sites, and the time the forest gets to restore, all contribute to the possible outcome of the succession process (Grubb 1977; Opler *et al.* 1977; Brokaw 1985a; Hubbell & Foster 1986, Huston & Smith 1987). Again, the chance of new seeds of a given plant species to arrive in the disturbed area depends on the timing of the disturbance (whether a plant species has viable seeds or advanced regeneration available at that moment), the spatial scale of the disturbance, coupled with the maximum dispersal distance of that plant species (Grubb 1977; Brokaw 1985; Opler *et al.* 1977). Animals can have a major effect on successional patterns through seed dispersal, seed-eating or herbivory (Terborgh 1992; Hammond & Brown in press). Species that show a rapid growth rate, and which can grow under high light environments, will be more important in early succession. The most rapidly growing plants tend to have the best dispersable seeds, based on small size and large numbers, and usually reach an open site before the seeds of the plants that will eventually predominate (Grubb 1977). However, during vegetation recovery the gaps will close, the abiotic and biotic conditions occurring in those gaps will chance, changing the capabilities of species to become dominant.

In this study the spatial component of logging damage, the location, size and form of gaps and skid trails, and the effects on the botanical diversity at different spatial scales will be emphasized. Furthermore, an evaluation will be made concerning vegetation recovery through time by comparing logged-over areas at different lengths of recovery time.

Measuring and monitoring botanical diversity

To be able to evaluate changes in botanical diversity as a result of logging it is necessary to measure it. So immediately the question arises 'What to measure' and 'How to measure it'.

In this thesis changes in plant species diversity (in richness and abundance) and functional diversity will be evaluated.

Species diversity

As already mentioned earlier, species diversity is usually expressed as a function of two parameters: species richness, which is number of species per unit area, and species abundance, which is number of individuals per unit area. Species richness is the most frequent and widely applied measure of biodiversity, and this contributes to the fact that it is usually considered as the fundamental point of focus of biodiversity (Gaston 1996). Data on species richness already exist through former (herbarium) collections and inventories, and it is relatively easily measured compared to *e.g.*, genetic variation within species. Although it represents only one level of biodiversity, there is evidence that differences in species richness are related to differences in several other measures of biodiversity. Species richness, for instance, may be positively related to habitat diversity (*e.g.*, Magurran 1988; Brewer & Williamson 1994), positively related to diversity of higher taxonomical units (Williams & Gaston 1994), related to topographic and edaphic diversity (*e.g.*, Gentry 1990; Primack 1993; Huston 1994; Clinebell 1995) and species richness captures some elements of functional diversity (Gaston 1996).

Indices that are based on the proportional abundances of species are the commonly used Simpson's Index and Shannon Index. These indices, however, are heavily dependent on sample size, combine richness with a measure of evenness of relative abundance, and reduce information content (Magurran 1988). The Simpson's Index is actually a dominance measure, because the index is weighted towards the abundances of the most frequent species in the sample; it is less sensitive to species richness. Especially in tropical rain forests, where 'common is rare and rare is common' (Hoogmoed 1992), the underlying species abundance distribution is important in determining whether the index has a high or low value.

In this thesis simply species per unit area, and abundance per unit area will be used to quantify diversity.

To compare species diversity measures from different areas, and at different spatial scales, usually three types of diversity measures are being used (Whittaker 1977): alpha-diversity, beta-diversity and gamma-diversity. Within one habitat, the number of plant species per area of a given size (usually < 1000 m^2) is a good first order estimate of local plant diversity (alpha diversity) (*e.g.*, Gentry 1990; Valencia *et al.* 1994; Clinebell *et al.* 1995; Condit *et al.* 1996), although rare species are expected to be under-represented. Where several such samples are repeated in adjacent sites that differ in environmental characteristics, they can be used to estimate the spatial turnover in species between those sites (beta diversity) (*e.g.*, Clinebell *et al.* 1995). Gamma-diversity used as defined by Whittaker (1977), would indicate a the number of species within a region. Cody (1986), however, uses gamma-diversity as a measure of differences in species composition between similar habitats in different geographical areas.

In this thesis all three types of diversity (alpha-, beta- and gamma-diversity (sensu Whittaker 1977) measures will be evaluated.

Functional diversity

Responses to environmental factors are likely to vary among species that participate in a particular ecosystem process, with some species increasing in abundance and others decreasing in response to a single gradient or chance in conditions. It is argued that biodiversity patterns can best be understood if groups of species that are reasonably consistent in their responses to the physical environment are distinguished (Huston & Gilbert 1996). To classify species into such functional groups imposes a degree of simplification which reduces information content, but reveals general (and specific) biodiversity patterns and facilitates predictions about forest processes. The number of functional groups with respect to any particular process is termed the diversity, or richness, in functional groups. Diversity within functional groups refers to the number of species included in the group.

For the study of tropical forests where richness of species in most communities is unknown and ecological role of many described species is poorly understood, no accepted classification of functional groups exists (Orians *et al.* 1996). One proposed example of functional groups in tropical forests is growth forms (Ewel & Bigelow 1996).

Many studies on tropical rain forest merely address the woody component of the forest, *i.e.* trees and large woody lianas, and even mostly those with a dbh of more than 10 cm (*e.g.*, Putz *et al.* 1984; Campbell *et al.* 1986; Gentry 1990). According to Bruenig (1996) 'Trees are the primary structural elements of the forest, which foresters manipulate, produce and preserve', and this has resulted in an emphasis on the tree component of the forest in relation to sustainability research. In such an approach 'sustainable logging' only indicates sustained production of commercially interesting timber species, and not maintenance of the characteristic species diversity of the forest. Such an approach also neglects the process of competition on the forest floor in which all ground-rooted growth forms are participating.

In this study all growth forms will be considered, including herbs, shrubs, palms, lianas, epiphytes, and trees. The reactions of different growth forms in relation to logging will be evaluated, and, within those growth forms, groups of species that show similar reactions to logging will be defined.

Lianas

One of the growth forms that may have a specific relation to logging is the lianas. The term 'lianas' is used as a synonym for several types of climbers, such as woody climbing plants that rely on other plants for support (Putz *et al.* 1984), woody hemiepiphytes, which rely on other woody plants for their support for only part of their life cycle (Benzing 1995), and herbaceous tendril climbing plants (Gentry & Dodson 1987). Relatively little is known about lianas. Both Gentry (1991) and Lowman & Nadkarni (1995) state in their overview of lianas and canopy ecology that this group is clearly under-represented in the large herbaria specialized in tropical floras.

Lianas, due to their growth form, can connect crowns of individual trees, close the canopy, stabilize the microclimate below and provide pathways for arboreal animals, such as rodents (Malcolm 1991), monkeys (Emmons & Gentry 1983), and sloths (Montgomery & Sunquist 1978). Lianas form an important food resource for several animals (*e.g.*, Gentry 1991; Hammond & Brown 1995; Galetti & Pedroni 1994; Morellato & Leitão-Filho 1996) with a distinctly different phenological pattern from trees and treelets (Morellato & Leitão-Filho 1996). They could therefore play a key-stone function in this ecosystem, producing *e.g.* fruits in periods when other fruit resources are scarce (Johns 1992).

In relation to disturbance and in particular to logging, the connecting ability of lianas may cause more trees to collapse, in any event forming a gap, and the resulting gaps are larger (Hammond & Brown 1991; Van der Hout 1996). Regenerating forest areas are often particularly rich in lianas due to their strong resprouting abilities (Putz 1984b; Appanah & Putz 1984) and many are capable of quick recovery after stem damage (Fisher & Ewers 1991). In general, a high density of lianas can be interpreted as a biological marker of past forest disturbance (Hegarty & Caballé 1991). Lianas are generally considered to be opportunistic plants, colonizing open sites, hold them for a period and then decline in density as trees become dominant. Trellises, *i.e.* supports to cling or climb on to, are important for long term establishment. Lianas can also influence the growth of tropical trees and their seedlings (Clark & Clark 1990; Putz 1984a; Putz *et al.* 1984). In view of the above, foresters consider lianas as a nuisance and pre-harvest liana cutting is part of many silvicultural treatments (Fox 1968; Appanah & Putz 1984; Putz 1991; Van der Hout 1996; Mason 1996).

In this study special emphasis will be given to the response of lianas to logging.

Epiphytes

Another growth form that plays a very special role in the forest ecosystem is the epiphytes. Epiphytes are found to influence processes in tropical forest ecosystems such as mineral cycling, water storage and pulse-wise release, and nutrient storage, where they intercept the nutrient cycling between the tree and soil-litter pools (Benzing 1989; Nadkarni 1988; Veneklaas *et al.* 1991). True epiphytes are herbs or woody plants that germinate on other woody plants (Benzing 1995). They grow mainly in the canopy of trees. In general, abiotic conditions in the canopy are typified by intense sunlight, wide amplitudes in relative humidity, severe water stress, and a small, pulse-supplied pool of nutrients compared with forest floor conditions (*e.g.*, Benzing 1989, 1990; Coxson 1991, 1992; Parker 1995). The environmental requirements of epiphytes are expected to be provided by a great number of different host tree species, although it has been documented that different epiphytic vegetations can be found in different lowland rain forest types (Sandford 1969; Johansson 1974; Ter Steege & Cornelissen 1989). In general, epiphytes are slow growing due to low water and nutrient availability within the canopy. Changes in the epiphytic community are, apart from the disappearance of those epiphytic individuals that grow on the harvested trees, expected to be very slow. Epiphytes are therefore not expected to show any change in species richness related to logging on the time scale considered in this study.

In this thesis the species richness of epiphytes in the untouched Greenheart forest will be compared to that of other forest types occurring in the Guianas.

Objectives and approach

The main objective of this study is to analyze the impact of disturbance, and in particular logging, on species diversity and functional diversity of Greenheart (*Chlorocardium rodiei*) dominated rain forest, and to contribute to a sustainable management plan for this forest type containing Guyana's main commercial timber species. Other objectives are to quantify botanical diversity in Greenheart forest in Guyana, and to evaluate various hypotheses on the maintenance of diversity. Apart from quantification of the species composition of undisturbed Greenheart forest, an evaluation will be made of the development of plant diversity of regenerating Greenheart forest through time. Special emphasis will be given to the spatial component of diversity, *i.e.* the location and relative abundance of species, in relation to the spatial component of logging damage: the location, size and form of gaps and skid trails. In this study groups of species that show similar spatial distributions patterns before and after logging will be defined, and related to reactions to logging.

The study has been conducted on the timber concession of Demerara Timbers Ltd. near Mabura Hill (5°13'N, 58°48'W). The concession lies between the Essequibo and Demerara Rivers in north-central Guyana, approximately 240 km from the coast. For this study an evaluation was made of all plants collected in the concession in previous years and during the fieldwork for this thesis (1992-1996). Within Greenheart forest, a total of 24 hectares were inventoried. According to Van Kekem *et al.* (1996), who made a detailed soil description of the whole concession, all established one-ha plots are situated in the same mapping unit: deep, well-drained yellowish brown to strong brown sands, often with sandy loam to clayey subsoils. The soils are poor in nutrients and have moderate amounts of available soil moisture capacities. Fifteen of the 24 one-ha plots are situated within the reduced-impact logging project, in the West Pibiri compartment (indicated in this thesis as 'Pibiri'). The enumerations in Pibiri took place during 1993 and 1994. During 1994 and 1995, three exploited sites of Greenheart forest were sampled: at 2 km south of Mabura township ('2 km') logged in 1985, in kilometer square no. 34 in the Waraputa compartment ('Waraputa') logged in 1989-1990, and just outside the Mabura Hill Forest Reserve ('Reserve') logged in 1989. In each of these exploited sites three one-ha plots were established, two logged and one referential. The latter three sites were included to investigate the influence of time after logging, and introduced another level of geographical differences. The exploited sites were all logged with comparable logging intensities and conventional logging techniques. In 1996, the lianas of two one-ha plots were re-enumerated in Pibiri two years after logging.

Biodiversity in relation to forest management cannot be assessed at the scale of a forest management unit alone, but must also be seen in a landscape and regional geographical setting. Botanical diversity of Guyana's Greenheart forest can be expressed as species diversity, growth form diversity, and diversity of functional units. The history of an area is reflected in the available pool of species within the region. The potential changes are also reflected in the available pool of species. After disturbance, the forest opens up and new, invading species may arrive. Invading species, however, need to be present somewhere in the surroundings of the logged areas in order to be able to colonize the area. The level of historical disturbance determines whether the forest and its surroundings still mainly have primary forest species, or already contain many introduced 'weedy' species. This leads to the following research questions:

1 Which plant species can be found in the Mabura Hill area?
2 Is the flora of the Mabura Hill area a taxonomically random assemblage from the available total flora of the Guianas?
3 Is there any indication of large-scale historical disturbances in the area?

These questions are addressed in **chapter** 2.

Plant diversity includes the species richness and abundance of trees, palms, herbs, shrubs, lianas, and epiphytes. Epiphytic richness is not expected to show any change related to logging on the time scale considered in this study. Nevertheless, richness in epiphytes contributes to a large extent to the species richness in the Neotropics (Gentry & Dodson 1987). Therefore, epiphytes are treated separately and the following questions were raised:

4 What are the patterns of epiphytic species richness in different rain forest types?
5 Are vertical distribution patterns of epiphytic species richness the same in different rain forest types?
6 Can functional groups, *i.e.* groups of species that show the same vertical distribution pattern or partition the same environment within a tree, be distinguished in different rain forest types?
7 Can differences in species richness between the forest types be related to geographical differences, forest type specific differences, or host-specific differences?

These questions are addressed in **chapter** 3.

To evaluate the effects of logging on the diversity of Greenheart forest it is necessary to first quantify and describe its botanical diversity, and then to compare the pre- and post-logging species richness and abundance.
This leads to the following research questions:

8 What are the patterns of species richness and abundance in Greenheart forest before logging?
9 What are these patterns of species richness and abundance in Greenheart forest at different points in time after logging?
10 Is there a relation between species richness and abundance at different spatial scales and geographical distance?
11 Is this relation the same before and after logging?
12 Is there a relation between species richness and abundance after logging and logging damage?
13 If yes, is this relation the same for different growth forms?
14 Is there a difference in species composition between different spatial units, *e.g.*, different zones in and around a gap, or small gaps (<300 m^2) and large gaps (≥ 300 m^2), between unlogged and logged plots?

These questions are addressed in **chapter** 4.

Regenerating forest areas are often particularly rich in lianas due to their strong resprouting abilities. A high density of lianas is considered a biological marker of forest disturbance. This would suggest that the reactions of lianas to logging are highly correlated to the disturbance. This leads to the following research questions:

15 Is logging correlated with a high level of species richness and abundance of lianas?
16 Can we distinguish an influence of logging outside the directly affected forest areas (such as gaps and skid trails)?
17 Can we distinguish preferential establishment sites for lianas?
18 Can we distinguish functional groups of liana species responding to the logging in a similar way?

These questions are addressed in **chapter 5**.

Logging will lead to a mosaic of patches of forest, all of the same successional stage. As a result of logging, and all processes described in this chapter, biodiversity changes in time and space. Temporal variability is manifest in the differences in time since logging between the study sites, and partly between logged and unlogged plots within a site. Spatial variability is apparent at several scales: the different geographical locations of the study sites, between and within plots, and the different spatial effects of logging. The effects of logging can therefore be evaluated within plots, between plots, within study sites and between study sites. The patterns found can be used to draw conclusions concerning the applicability of theories concerning maintenance of species diversity for a tropical rain forest in Guyana, and to draw conclusions concerning the implications for forest management.

Concerning theories of maintenance of diversity the following question can be asked:

19 What conclusions can be drawn with regards to the different theories concerning the maintenance of diversity?

This question is addressed in **chapter 6**.

The implications of the obtained results for sustainable management of a tropical rain forest, with recommendations for forest management, are discussed in **chapter 7**.

The summary is presented in **chapter 8**.

CHAPTER 2

The flora of the Mabura Hill area, Guyana

Renske C. Ek & Hans Ter Steege

Abstract

This paper describes and analyses the composition of the flora of the Mabura Hill area, Guyana, South America. In total a number of 1433 different vascular plant species were collected in the vicinity of Mabura Hill. This flora is compared with the flora of other Neotropical sites, and in particular with the flora of other sites from the Guayana Shield. Using a Monte Carlo simulation technique, we evaluated the pattern of occurrences of plant families. Species (and consequently families) occurring in the Mabura Hill area are not a taxonomically random assemblage from the available total flora of the Guianas. Over-represented plant families contain mainly primary forest elements, while the under-represented families indicate the paucity of undergrowth species (herbs, shrubs, and small palms) and epiphytic species, especially when compared to two French Guianan sites.

The comparison shows that the Mabura Hill area is well representative of lowland Guianan rain forest, with little large-scale natural disturbance.

Introduction

Guyana is situated in NE South America, between Suriname, Venezuela, and Brazil. As a part of the Guianas (Guyana, Suriname, and French Guiana), the country belongs to a distinct phytogeographical province on the Guayana Shield, which also covers a part of Venezuela and Brazil. This phytogeographical province can be subdivided in four floristic regions (Berry *et al.* 1995): 1. Eastern Guayana Province; 2. Central Guayana Province; 3. Western Guayana Province; 4. Pantepui Province. Within those floristic regions, the Guianas are mostly part of the Eastern Guayana Province. Part of Guyana, the Pakaraima Mountains area, however, belongs to the Central Guayana Province, while the Mt. Roraima area belongs to the Pantepui Province.

A proper understanding of the biodiversity of the tropical rain forests of the Guianas is essential for the wise utilization of these forests, and for efforts to conserve as much as possible of the biological richness of these forests. Despite the breadth of the concept of biodiversity, number of species (or species diversity) remains its most frequently and widely applied measure. Species diversity is an easily measurable parameter. Moreover, quite some data on species richness already exist for most geographical areas through former botanical collections. Together the three Guianas are estimated to possess between 7.000 and 10.000 species of vascular plant species (Lindeman & Mori 1989), and, using the former botanical collections, almost 9000 species are described for the area at present (Boggan *et al.* 1992; with additions from Lindeman, pers. comm.). Still, since Davis & Richards (1932, 1934) published on the botanical inventory plots at Moraballi Creek, Guyana, only few quantified data about species diversity of the Guianan region has been produced (Ter Steege *et al.* 1993; Comiskey *et al.* 1994; Johnston & Gillman 1995). On the scale of worldwide biogeographical patterns, the similarity in the floras of South America and Africa, is well documented (*e.g.* Goldblatt 1993). Both continents originated from West Gondwanaland, and according to Pitman *et al.* (1993) the final separation took place between Western Africa and Northern South America (Northern Brazil and the Guianas). The Guayana Shield and Western Africa are part of the same ancient geological shield, and neither area has known any major geological chance since (Fjeldså & Lovett 1997) in contrast to other parts of South America which are more recent in origin (Pitman *et al.* 1993). There is evidence for both areas concerning later periodic aridity alternating with periods of raised lake levels in the Pleistocene (Colinvaux 1993; deMenocal 1995). We hypothesize that this common ancestry and comparable selection pressures since is especially noticeable in similarity between the flora of the Guayana Shield and that of Western Africa, especially for 'old species', with lineages of pre-Pleistocene age (after Fjeldså & Lovett 1997).

In order to be able to compare within and between floristic regions of the Guayana Shield phytogeographical region, three sites in French Guiana (Eastern Guayana Province), one site in Venezuela, and one site in Colombia (Western Guyana Province) are compared with Mabura Hill (Eastern Guyana Province). Furthermore, other Neotropical sites are taken into account, that do not belong to the Guayana Shield phytogeographical region.

In order to compare the composition of the flora of Mabura Hill with larger (worldwide) biogeographical patterns of diversity, we evaluate the origin of the most important plant families. Patterns of diversity are compared with those from Africa.

The aim of this study is to describe and analyse the composition of the flora of the Mabura Hill area, in terms of taxonomical composition and growth form spectrum, and to compare this flora with other Neotropical sites, with particular reference to other sites on the Guayana Shield.

Methods

Study area

The study area consists of approximately 10.000 ha of lowland forest in the vicinity of Mabura Hill (5°13'N, 58°48'W), between the Essequibo and Demerara Rivers. In general the area is gently undulating, and notwithstanding the fact that it belongs to the White Sands area (Daniel & Hons 1984), it occasionally is penetrated by laterite-covered dolerite dykes from the Pre-Cambrian Guayana Plateau, forming ridges and hills. The White Sands area, with its white sands, brown sands, and loams (Van Kekem *et al.* 1996), has a drainage pattern of small creeks. Where the white sand layers are thin, as in part of the study area, the streams have cut through this layer into the brown (loamy) sands and sandy clay loams below. The soils in the area are summarized in Van Kekem *et al.*(1996).

The climate is tropical with high rainfall, between 2500 and 3400 mm per year, and an average of 2700 mm. There are two distinct dry periods: January to March and August to September. There is no month, however, with an average rainfall of less than 100 mm. Mean daily average temperature over the year is 25.9 °C (Ter Steege *et al.* 1996).

The tropical rain forest of Guyana is peculiar in the sense that large parts are dominated by a few or one species (Davis & Richards 1934; Fanshawe 1952; Ter Steege *et al.* 1993). This dominance results in clearly identifiable forest patches that are often associated with particular soil types or conditions. The main 'forest types' represented in the inventory of the Mabura Hill area are summarized in Appendix 1. Further information can be found in Ter Steege *et al.* (1993, 1996).

Collection history and species lists

The oldest herbarium collections known from the Mabura Hill area, and incorporated in the species list, are from Jenman (1881-1898), who visited the area on several occasions (Ek 1990). Richard and Robert Schomburgk visited the area earlier, but as their collections were not numbered in chronological order and were only labeled 'Essequibo', these could not be included. Other collectors (and collecting institutions) for the area are: A.A. Abraham (1919), Louis Hohenkerk (1909-1923),

John Edgar Beckett (1906), Forest Department (1935), Paul Maas *et al.* (1981, 1988), Ton Stoffers *et al.* (1982), Hans Ter Steege *et al.* (1985-1992), John Pipoly *et al.* (1986-1987), Claude Persaud (1987), Marion Jansen-Jacobs *et al.* (1989), Bill Hahn (1989), Marcel Polak *et al.* (1990-1992), Renske Ek *et al.* (1992-1996), Helen Kennedy (1992), Eric Christensen & Carol Kellof (1994), David Clark (1996), and Sam Roberts (1996). All collections are present in the herbarium of the University of Guyana (BRG), and subsets available in Utrecht (U) and Washington (US). Since 1992 a small reference herbarium was established in Mabura Hill, and collections made after this date are also available in that herbarium.

In addition to the above-mentioned herbarium collections, species recorded in botanical studies and surveys have been added. These studies and surveys include: epiphytes from three forest types (Ter Steege & Cornelissen 1989), epiphytes from Mora forest (Biesmeijer & Bleij unpubl.), Great Falls Inventory (Welch & Bell 1971); Waraputa Compartment (Ter Steege *et al.* 1993; Ek *et al.* unpubl.); Pibiri Compartment (Van der Hout unpubl.; Ek *et al.* unpubl.), and inventory results from the Mabura Hill Forest Reserve and near '2km' (Ek *et al.* unpubl.). Furthermore, for Orchidaceae all species collected and kept as living material in Mabura Hill (*HtS & RE*) are incorporated.

Nomenclature follows the system of Cronquist (1981), with a few alterations after Maas *et al.* (1993). Due to the fact that familial limits are not agreed upon among ferns and fern allies, we follow the common practice to treat the group as one family (Gentry 1990). The three leguminous families are treated as one family group, Leguminosae, although we also review the contribution of the three separate sub-families.

A summary of the vegetation types at Mabura Hill is given in Appendix 1. Vegetation types 7, 9, 10, 11 (appendix 1) have not been collected as intensively as the other ones; species from these types can be under-represented in the sample.

The species composition of the flora of the Mabura Hill area is furthermore compared with four Western Amazonian and Central American sites (La Selva (Hartshorn & Hammel 1994, Wilbur *et al.* 1994), Barro Colorado Island (Hubbell & Foster 1990; Gentry 1993), Araracuara (Duivenvoorden & Lips 1995), and Manu (Foster 1990)) and six Eastern Amazonian sites (San Carlos de Río Negro (Clark, unpubl. 1996), Ilha de Maraca (Milliken & Ratter 1989), Saül (Mori *et al.*, pers. comm.), St. Elie (only trees) (Lescure *et al.* 1990), Nouragues (Poncy *et al.* 1996), and Reserva Ducke (Prance 1990, Gentry 1993).

Data analyses

To compare the contribution of the different plant families to the flora of Mabura Hill in relation to the complete known flora of the Guianas, we evaluated the random or non-random pattern of occurrences of plant families using a simple Monte Carlo simulation technique (Besag & Clifford 1989). The composition of the flora at Mabura Hill was tested against 50 random selections of 1400 species out of the total set of the 8708 species known for the Guianas. This latter list was obtained by using the 'Checklist of plants of the Guianas' (Boggan *et al.* 1992), augmented with data compiled from taxonomical literature and herbarium vouchers (Lindeman, pers. comm.). The average of these 50 runs was used as a randomized list to compare with the Mabura Hill flora. Differences between the randomized and observed species list were tested using Pearson chi-square tests. The same technique was applied to the species lists of two other sites in the Guianas, Saül and Nouragues, both in French Guiana.

To test differences in family composition between sites and between groups of sites, we assigned ranks per site to the most important families. Differences in ranks were tested with Kruskal Wallis one-way analysis of variance, using Statistica 4.5 (StatSoft, 1993).

Growth forms

All species within the species lists were, as much as possible, assigned to a growth form. The growth forms used are mainly defined after Hubbell & Foster (1990, 1992), although some modifications are made following Gentry & Dodson (1987). Species not taken into account are those which show polymorphism to such a degree that they are hard to classify (*e.g. Connarus coriacea* is listed as a tree in the literature (Forero 1983), but found in our collections four times as a tree and four times as a liana; subsequent herbarium search in NY, US, and MO showed the same pattern), or those of which the identification level impedes a classification.

Trees: The tree component includes the treelets, woody freestanding plants with an adult height between 4 and 10 m and a diameter > 10 cm, midstorey trees with an adult height between 10 and 20 m and canopy trees with a height above 20 m. Large palms, occurring in the mentioned size classes, are also included in this category.

The climbers; lianas, woody hemiepiphytes, and vines: Lianas are woody climbing plants that rely on other plants for support (Putz 1984); they are an important structural component of tropical forests. Woody hemiepiphytes rely on other woody plants for their support only for a part of their life cycle. They germinate either on the ground and then climb up, loosing their contact with the soil in a later stage (mostly Araceae) or germinate in the tree and send roots downwards. The latter group eventually can prevent the host tree from lateral growth (stranglers, like *Clusia* spp. and *Ficus* spp.) and it can take over part of the canopy or even the whole position of the original host tree. Vines are herbaceous, small tendril-climbing plants. For a matter of convenience the word lianas hereafter used refers to the whole group of climbers.

The undergrowth; herbs, shrubs, and small palms: Although neglected in research, the undergrowth generally contains between 25 to 46 % of the species in wet tropical forests (Gentry & Dodson 1987). In this study herbs are defined as terrestrial herbaceous plants and terrestrial woody plants with a diameter < 1 cm. Shrubs are woody, freestanding plants with an adult height < 4 m and with a diameter between 1 and 10 cm. Small palms are palms with an adult height of less than 4 m.

Results and discussion

Taxonomic richness

A total number of 1433 vascular plant species has been collected (until June 1996) at Mabura Hill, belonging to 120 plant families. A total of 1305 taxa (91.1%) could be identified to species level and 128 (8.9%) only to the level of genus. For a total list of species, and their authorities, see Appendix 2.

Families

The largest family group, Leguminosae, has a total of 132 species (Caesalpinoideae 39 spp., Mimosoideae 26 spp., and Papilionoideae 67 spp.) (Fig. 1). The second largest family is Orchidaceae with a total of 126 spp., followed by Rubiaceae (69 spp.).

Families significantly differing from the randomized species set resulting from the Monte Carlo simulations (N=50) are listed in table 1. Families significantly over-represented (ratio > 1) in the Mabura Hill flora compared with the randomized set are typical primary forest families, and especially lianas (10 out of 21). Eight of those families (marked with # in table 1) also belong to the top ten richest families of Mabura Hill. Lianas are often under-collected at many sites (Gentry 1991; Lowman & Nadkarni 1995). The emphasis of one of us (RE) on lianas could be reflected in the large proportion of over-represented liana families for the Mabura Hill area. Especially Loganiaceae, Dilleniaceae, Smilacaceae, Menispermaceae, and Connaraceae, all (mainly) climbing families, are found much more than expected. Over-represented tree families contain families listed as specific Guianan lowland families, such as Annonaceae, Chrysobalanaceae, Lecythidaceae, and Sapotaceae (Prance 1982; Mori 1991). Six of the over-represented tree families (marked with * in table 1) comprise large-seeded species, with fresh seed weights of > 5 g (Hammond & Brown 1995). According to those authors the occurrence of large seeds is a specific phenomenon of the flora of the Mabura Hill area, and indicates a low disturbance of the area.

Families significantly less represented (ratio < 1) contain mostly savanna- and disturbance species. This is in correspondence with the lowland forest nature of our sample.

We obtained significant similarities and differences performing the same Monte Carlo analyses for two other such sites in French Guiana for which complete checklists were available, namely Saül (Cremers *et al.* 1993; Mori *et al.* pers. comm), and Nouragues (Poncy *et al.* 1996), both standardized for 1400 spp. (table 1). The four specific Guianan lowland families also occur significantly over-represented at the other two sites (except Annonaceae for Saül), but lianas are relatively scarce. Compared with the randomized set they both have their own specific set of over-represented families. Tree families such as Moraceae, Myristicaceae, Meliaceae, and Anacardiaceae are all more prominently present at the French Guiana sites than in the Mabura Hill area. These families provide many non-timber forest products and

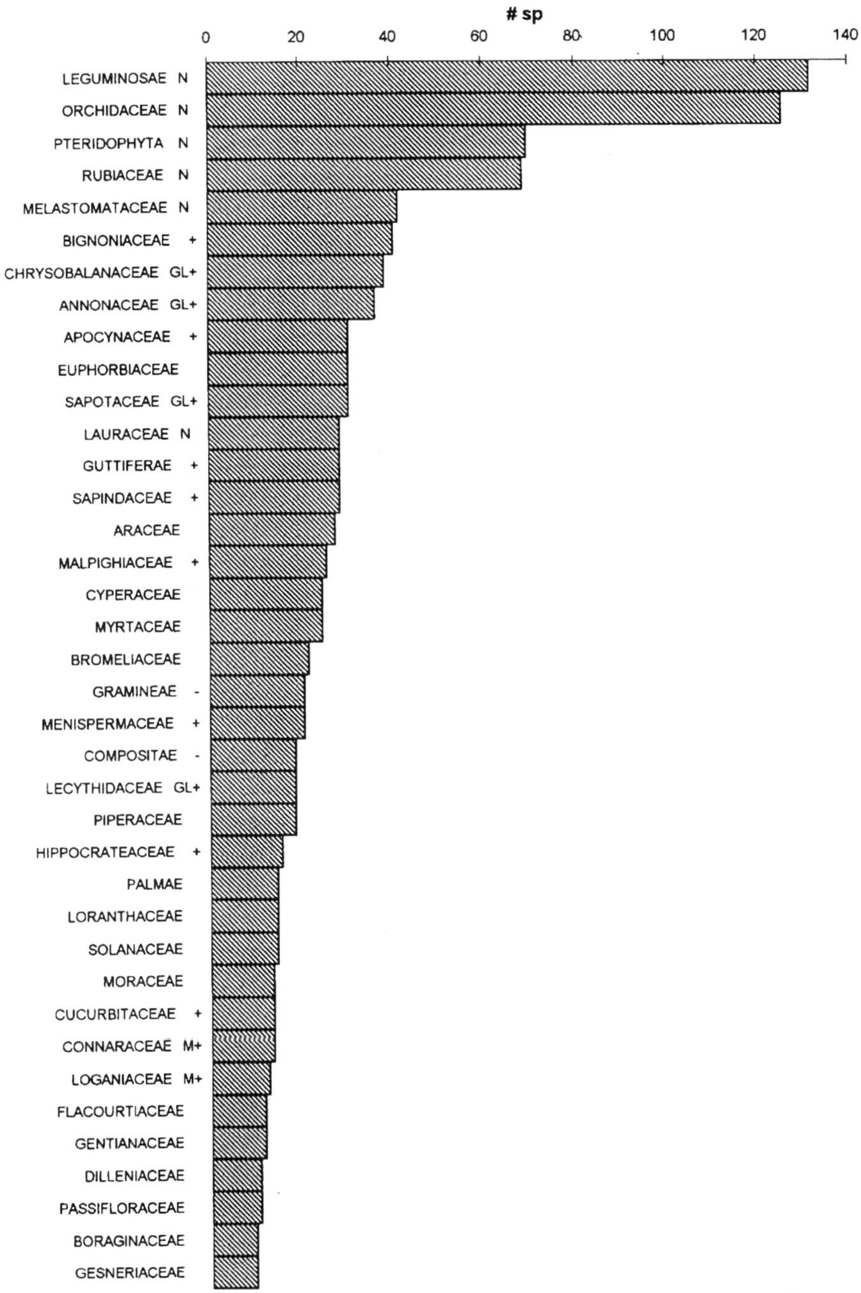

Figure 2.1

Families occurring in the Mabura Hill area, Guyana, with more than 10 spp. N= Neotropical (Gentry 1990), GL = Guianan Lowland (Huber 1995, Prance 1982, Mori 1991), M= Mabura (this study), +/- = more or less than expected (derived from table 1).

Table 2.1.
Comparison between randomized family distribution for the Guianas and three lowland Guianan sites. Ratio observed/random only given when significant (Pearson Chi-square test, p< 0.05).

		random (50 trials)		Mabura Hill		Saül		Nouragues	
Family	Type	mean	stdev	N	ratio	N	ratio	N	ratio
Gramineae	S/D/U	57	6.95	21	0.37	26	0.46	25	0.44
Malvaceae	S/D/U	10	2.43	4	0.40	12		2	0.23
Asclepiadaceae	D	13	3.48	5	0.40	3	0.21	2	0.21
Acanthaceae	S/H/U	14	3.87	6	0.43	26	1.83	2	0.16
Ochnaceae	H/U	10	2.80	6	0.60	7		11	
Cyperaceae	D/U	42	6.25	25	0.60	18	0.43	27	0.67
Compositae	H/D/U	29	3.96	19	0.66	26		6	0.19
Orchidaceae #	E	102	7.06	126	1.24	101		77	0.76
Apocynaceae #	L/T	23	3.94	31	1.35	26		26	
Guttiferae # *	L (HE)	20	3.92	29	1.45	18		23	
Lauraceae *	T	18	4.04	29	1.58	29	1.63	24	
Sapindaceae #	T	18	3.17	29	1.63	14		7	0.45
Gentianaceae	U	7	3.79	12	1.71	36	1.98	27	1.47
Malpighiaceae	L	15	4.02	26	1.73	38	2.16	66	3.88
Sapotaceae # *	T	17	2.36	31	1.79	11		14	1.95
Annonaceae #	T	20	4.04	37	1.85	27		31	1.57
Bignoniaceae #	L	21	3.27	41	1.97	29	1.39	16	
Cucurbitaceae	L	7	2.80	14	2.00			7	
Chrysobalanaceae # *	T	19	4.18	39	2.05	23		41	2.15
Lecythidaceae	T	8	2.35	19	2.29	22	2.69	30	3.48
Loganiaceae *	L	5	2.11	13	2.40	11	1.94	8	
Hippocrateaceae	L	6	2.62	16	2.67	9		15	2.27
Dilleniaceae	L	4	2.00	11	2.75	3		5	
Smilacaceae	L	3	1.51	9	3.00	3		6	2.84
Simaroubaceae *		2	1.37	6	3.00	10		7	
Menispermaceae	L	6	2.47	21	3.50	4		6	2.84
Nyctaginaceae		2	1.15	7	3.50	3		6	

Table 2.1. (continued)

Families are ordered to ratio less than expected to more than expected, first for Mabura Hill, afterwards Saül and Nouragues. Latter sites standardized for 1400 spp.

		random (50 trials)		Mabura Hill		Saül		Nouragues	
Family	Type	mean	stdev	N	ratio	N	ratio	N	ratio
Connaraceae	L	4	1.79	14	3.50	4		2	
Rubiaceae #	U	73	8.35	69		102	1.40	72	
Solanaceae	U	13	2.85	15		20	1.50	18	
Piperaceae	U	22	4.57	19		35	1.59	38	1.79
Marantaceae	U	10	2.73	5		17	1.64	7	
Araceae	E	22	3.94	28		39	1.78	51	2.38
Moraceae		16	3.27	14		29	1.81	56	3.48
Myristicaceae	T	3	1.59	6		5	1.91	8	3.98
Meliaceae *	T	6	2.04	7		12	2.05	18	3.03
Pteridophyta #	U	74	7.28	70		156	2.12	**	
Urticaceae	U	2	1.46	1		5	2.21	1	
Burseraceae *	T	7	2.64	9		18	2.48	28	4.06
Ebenaceae	T	4	1.72	4		9	2.50	5	
Musaceae	U	2	1.47	5		7	2.92	5	
Anacardiaceae	T	3	1.81	4		8	3.13	10	5.11
Convolvulaceae	L	13	3.14	9		8		5	0.35
Palmae *	T	16	3.85	15		22		30	1.85
Passifloraceae	L	8	2.69	11		8		18	2.27
Mimosoideae (Leg.)	T	28	4.68	26		**		68	2.44
Lentibulariaceae	U	5	2.35	1		0		13	2.50
Burmanniaceae	U	3	1.48	6		6		8	2.65
Gesneriaceae	U	7	2.98	10		13		19	2.76
Celastraceae		2	1.25	2		3		6	2.84
Monimiaceae	U	2	1.14	3		3		6	2.84
Cyclanthaceae	U	3	1.63	4		5		9	3.03
Olacaceae *	T	2	1.39	5		5		8	3.98

Types: D = disturbance, E= epiphyte, H= herb, HE= hemiepiphyte, L= liana, S= savanna, T= tree, U= undergrowth. #: one of the ten richest families for the Mabura Hill area; *: containing large seeded (> 5 g) species (Hammond & Brown 1995).

Table 2.2
Comparison of the (ten) richest families of the Mabura Hill area with a) five Neotropical sites outside the Guayana Shield, and b) five Neotropical sites on the Guayana Shield.

A

Rank	Mabura Hill	Ducke	La Selva	BCI	Manu	Ilha de Maraca
1	Leguminosae	Leguminosae	Pteridophyta	Pteridophyta	Leguminosae	Leguminosae
2	Orchidaceae	Sapotaceae	Orchidaceae	Leguminosae	Moraceae	Rubiaceae
3	Rubiaceae	Rubiaceae	Araceae	Orchidaceae	Rubiaceae	Cyperaceae
4	Pteridophyta	Chrysobalanaceae	Rubiaceae	Rubiaceae	Pteridophyta	Pteridophyta
5	Melastomataceae	Lauraceae	Melastomataceae	Araceae	Orchidaceae	Gramineae
6	Chrysobalanaceae	Annonaceae	Leguminosae	Moraceae	Acanthaceae	Moraceae
			Piperaceae			
7	Bignoniaceae	Moraceae	Euphorbiaceae	Bignoniaceae	Melastomataceae	Bignoniaceae
	Annonaceae		Moraceae	Piperaceae	Sapindaceae	Palmae
			Palmae			
8	Apocynaceae	Lecythidaceae	Lauraceae	Melastomataceae	Bignoniaceae	Orchidaceae
						Euphorbiaceae
9	Sapotaceae	Melastomataceae	Acanthaceae	Sapindaceae	Araceae	Myrtaceae
			Gesneriaceae			
10	Guttiferae	Palmae	Bignoniaceae	Apocynaceae	Myrtaceae	
	Euphorbiaceae				Annonaceae	
					Apocynaceae	
					Sapindaceae	
	120 families	88 families	121 families	118 families	130 families	118 families
	1433 spp.	825 spp.	1668 spp.	1320 spp.	1856 spp.	1100 spp.
Lit:	ter Steege et al. 1996 this study	Gentry 1990	Hartshorn & Hammel 1994	Gentry 1990	Gentry 1990	Milliken & Ratter 1989

are often planted or favoured in agroforestry (Prance & Kallunki 1984; Van Andel pers. comm.). This may indicate a long history of Amerindian settlements in the Nouragues and Saül area affecting our results. Furthermore, Nouragues has a striking dominance of Leguminosae (Mimosoideae), with many *Inga* species, some of which are well-known successional species (Schulz 1960). This high percentage may well be partly due to the research interest of one of the main investigators at that site, O. Poncy. Other over-represented families for the two sites are mainly undergrowth families, herbs and shrubs (17 out of 20). Although these over-represented families for the two French Guianan sites do occur at Mabura Hill, the results suggest that Mabura Hill has a relatively low diversity in undergrowth species. Although lowland rain forests of the Guianas are considered to be floristically similar (sensu

B

Rank	Mabura Hill	San Carlos de Río Negro	Araracuara	Saül	St. Elie (only trees)	Nouragues
1	Leguminosae	Leguminosae	Leguminosae	Pteridophyta	Leguminosae	Leguminosae
2	Orchidaceae	Rubiaceae	Annonaceae	Leguminosae	Lecythidaceae	Orchidaceae
3	Rubiaceae	Melastomataceae	Pteridophyta	Rubiaceae	Chrysobalanaceae	Rubiaceae
4	Pteridophyta	Apocynaceae	Rubiaceae	Orchidaceae	Sapotaceae Annonaceae	Sapotaceae
5	Melastomataceae	Cyperaceae	Melastomataceae	Melastomataceae	Moraceae	Moraceae
6	Chrysobalanaceae	Euphorbiaceae Palmae	Sapotaceae	Araceae	Apocynaceae Guttiferae Melastomataceae	Araceae
7	Bignoniaceae Annonaceae	Chrysobalanaceae	Euphorbiaceae	Sapotaceae	Euphorbiaceae Lauraceae	Melastomataceae
8	Apocynaceae	Araceae	Burseraceae	Lauraceae	Burseraceae Rubiaceae Sapindaceae	Chrysobalanaceae
9	Sapotaceae	Sapotaceae	Guttiferae Moraceae	Piperaceae Euphorbiaceae	Sterculiaceae Myrtaceae Flacourtiaceae	Myrtaceae Piperaceae
10	Guttiferae Euphorbiaceae	Gramineae Lauraceae	Palmae	Sapindaceae Moraceae Bignoniaceae		Annonaceae Lecythidaceae
	120 families 1433 spp.	105 families 812 spp.	** **	143 families 2314 spp.	49 families 288 spp.	120 families 985 spp.
Lit:	ter Steege et al. 1996 this study	Clark (pers. comm 1996)	Duivenvoorden & Lips 1995	Cremers et al. 1993 S. Mori et al. pers. comm.	Lescure et al. 1990	Poncy et al. 1996

Tolmatschew 1971; Berry *et al.* 1995), the three investigated sites differ to quite a degree.

The data of Mabura Hill were furthermore compared with four Western Amazonian and Central American sites (La Selva, Barro Colorado Island, Manu, and Araracuara) and six Eastern Amazonian sites (San Carlos de Río Negro, Ilha de Maraca, Saül, St. Elie, Nouragues, and Reserva Ducke) (table 2). Of those sites Araracuara, San Carlos de Río Negro, Saül, Nouragues, and St. Elie also belong to the Guayana Shield.

A total of 64 families occur at all sites. A high species diversity of Leguminosae is typical of all lowland Neotropical forests (Gentry 1988, 1993), and in our comparison

Table 2.3
Kruskall-Wallis analyses of variance by ranks between three phytogeographical regions in South America.

	Guayana		E. Amazonia		W. Amazonia & C. America	
CAESALPINOIDEAE	7.41	**	1.91	ns	6.25	*
APOCYNACEAE	3.16	ns	3.16	ns	3.16	ns
ANNONACEAE	1.41	ns	0.20	ns	1.41	ns
SAPOTACEAE	0	ns	1.40	ns	0.00	ns
CHRYSOBALANACEAE	0	ns	0.50	ns	0.00	ns
LECYTHIDACEAE	0	ns	0.00	ns	0.00	ns

this family is ranked between 1-6, with a median of rank 1. Among the Leguminosae, Caesalpinoideae are significantly more common at the sites belonging to the Guayana Shield (table 3) and the Brazilian sites (Kruskal-Wallis ANOVA on ranks, $H=7.41$, $p<0.01$), compared to Western Amazonian and Central American sites (Kruskal-Wallis ANOVA on ranks, $H=6.25$, $p<0.05$). The Eastern Amazonian sites do not show a significant correlation with the amount of Caesalpinoideae (table 3). The percentage of Caesalpinoideae/Leguminosae for the Guayana Shield Sites are: San Carlos de Río Negro 49%, Araracuara 33%, Mabura Hill 30%, Saül 28%, St. Elie 28%, Nouragues 23%, (the Guianas in total 30%); and for the Brazilian sites: Bahia 26% (Lewis 1987), Ducke 27%. For the Western Amazonian and Central American sites these ratios are: Manu 16%, Barro Colorado Island 14 %, and La Selva 13%. Remarkable is the low proportion of Caesalpinoideae at Ilha de Maraca in Brazil (17%), geographically close to Guyana, but just off the Guayana Shield into the Amazon basin. It must be noted, however, that the taxonomic placement of the genus *Swartzia* in either Papilionoideae (Polhill & Raven 1981; this study), or Caesalpinoideae (Cowan & Lindeman 1983) is debatable. Placement of *Swartzia* within the Caesalpinoideae would even enlarge the proportion of that family. This latter notion applies especially in the Guayana Shield sites, due to the biogeographical distribution pattern of *Swartzia* with its centre of distribution in northern South America.

Families such as Leguminosae, Orchidaceae, Rubiaceae, Lauraceae, and Melastomataceae are typically among the top twenty at all Neotropical sites and especially the first three belong to the most species-rich families all over the world (Heywood 1985). Orchidaceae are missing from the lists for the following sites: Reserva Ducke, Manu, San Carlos de Río Negro, and St. Elie. This could indicate that the species lists for those sites have an emphasis on the woody components of the forest. One family, Moraceae, appears in the top twenty for all other sites except San Carlos de Río Negro and is missing in the top twenty for Mabura Hill. Compared with the randomized set, however, the difference is not significant.

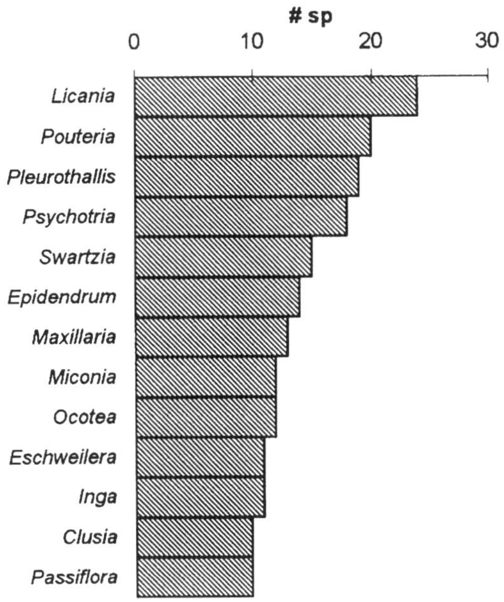

Figure 2.2
Largest genera Mabura Hill area, Guyana, with more than 10 spp.

Other families are listed as typically dominant (in species diversity) in Eastern Amazonia and thus in the Mabura Hill area. Such families include Annonaceae, Chrysobalanaceae, Lecythidaceae, and Sapotaceae (Prance 1982; Mori 1991), which occur mostly in lowland forest. These families, however, were not found significantly related to Eastern Amazonian sites using Spearman rank correlations (table 3).

Connaraceae, so well represented in the Mabura Hill area with 13 spp., are hardly found at other sites (Araracuara 1 sp., La Selva 2 spp., BCI 4 spp., Manu (not listed, if present < 5 spp.), Reserva Ducke 3 spp., Ilha de Maraca 2 spp., San Carlos de Río Negro 1 sp., Nouragues 2 spp., Saül 6 spp.).

Genera

The largest genera found in the flora of the Mabura Hill area are mostly tree genera and epiphytic genera (Fig. 2). *Licania, Pouteria, Swartzia, Inga, Ocotea,* and *Eschweilera* are mainly trees or treelets, *Psychotria* and *Miconia* are treelets or shrubs, and *Pleurothallis, Epidendrum,* and *Maxillaria* are large genera of epiphytic Orchidaceae. All genera (except *Eschweilera*) belong to the largest families listed above and are speciose throughout most of the Neotropics.

Genera occurring in the Mabura hill area and largely restricted to the Guayana Shield except for one or more outliers are: *Dicymbe* (Leguminosae), *Macrocentrum* (Melastomataceae), and *Thurnia* (Thurniaceae). Three genera are found only in the Guayana Shield floristic area; *Potamoganos* (Bignoniaceae), *Maburea* (Olacaceae), and *Cheiradenia* (Orchidaceae).

Species

The following new species were collected in the course of this inventarisation: *Maburea trinervis* Maas (Olacaceae) and *Salacia maburensis* Mennega (Hippocrateaceae). Furthermore *Ocotea* sp. nov. (Lauraceae), *Hymenolobium sp.* nov., and *Swartzia* sp. nov. (Leguminosae), and *Solanum* sp. nov. (Solanaceae) were collected.

In the flora of the Mabura Hill area there are quite a number of specific Guayana Shield elements. Species listed as endemic for the Guayana shield (de Granville 1988; Berry *et al.* 1995, Gentry 1990) or inadequately collected in other areas are: *Bactris ptariana* and *Jessenia bataua* subsp. *oligocarpa* (Palmae), *Potamoganos microcalyx* (Bignoniaceae), *Alexa leiopetala, Chamaecrista adiantifolia* subsp. *pteridophylla, Chamaecrista apoucouita, Eperua* spp., *Mora gonggrijpii, Swartzia roraimae,* and *Vouacapoua macropetala* (Leguminosae), *Dalechampia olympiana* (Euphorbiaceae), *Eriocaulon heterodoxum* (Eriocaulaceae), *Chlorocardium rodiei* and *Endlicheria chalisea* (Lauraceae), *Couratari gloriosa* (Lecythidaceae), *Quekettia microscopica* and *Cheiradenia imthurnii* (Orchidaceae), *Guzmania lingulata* (Bromeliaceae), *Anthurium jenmanii* (Araceae), *Passiflora cirrhiflora, P. fuchsiiflora, P. garckei,* and *P. laurifolia* (Passifloraceae), *Licania majuscula, Parinari campestris, P. parvifolia,* and *P. rodolphii* (Chrysobalanaceae). Species limited to the Guianas and part of Brazil are *Passiflora glandulosa* (Passifloraceae) and *Licania heteromorpha* (Chrysobalanaceae).

Only three introduced 'weedy' species were found (*Citrillus lanatus* and *Lagenaria siceraria* (Cucurbitaceae), and *Vigna luteola* (Leguminosae). Boggan et al. (1992) also cited a level of introduced and naturalized species for the whole of the Guianas as only 3 %. This low number suggests that most of the area has little large-scale disturbance.

Growth forms

Of the 1433 species, a total of 1388 (98 %) could be assigned to growth form (Fig. 3). The categories used are ordered from the forest floor to the canopy, that is herbs, shrubs, epiphytes, lianas, and trees.

Mabura Hill has a high diversity in liana species compared to other sites. Percentage of epiphytes and herbs, however, is lower than at other sites.

According to Hartshorn & Hammel (1994) a high species diversity of herbs is correlated with wetness of the forest (soil moisture), due to the fact that forests on wetter soils are mostly more dynamic (Phillips *et al.* 1994) and therefore more frequently illuminated. Furthermore, a high number of epiphytes is in general associated with climatologically wetter areas (air moisture) (Gentry 1990; Benzing 1995). The largest portion of the epiphytes found in Mabura Hill were small 'exposure type' orchids (sensu Benzing 1995). Exposure type epiphytes grow (mostly) in the canopy, exhibit a wide tolerance to exposure to light and can endure prolonged periods of drought (Benzing 1995). The occurrence of mainly exposure type epiphytes, paired with low species diversity of herbs, and scarcity of ferns (and

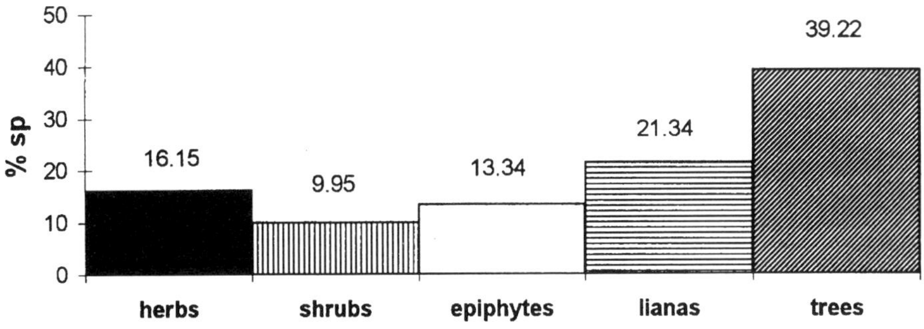

Figure 2.3
Growth form spectrum of species found in the Mabura Hill area, Guyana.

especially filmy ferns (Hymenophyllaceae) that can occur on lower, wetter, parts of the tree), suggest that the forest in the Mabura Hill area is a relatively dry forest.

The annual amount of rainfall, however, is sufficient for tropical rain forest. Monthly rainfall is never less than 100 mm (Van Kekem *et al.* 1996). Jetten (1994), however, indicates that within the white sandy soils, where the main body of data has been gathered, low water retention capacity may occur, which might indicate that there is repeatedly water limitation for the plants.

A preliminary comparison with inventory data of the Forest Department of Guyana, seems to indicate that a low species diversity is found for all sites in the White Sands Plateau (Berbice or Zanderij formation) (Ter Steege, unpubl. data), confirming the notion on the possible role of the soil type. Moreover, two vegetation types on white sand occurring in the area (appendix 1), locally known as 'Wallaba forest' and 'Muri shrub', are ranked among the most species-poor forests in South America (Eiten 1992).

Biogeographical affinities

In order to compare the composition of the flora of Mabura Hill with larger (worldwide) biogeographical patterns of diversity, we evaluated the probable origin of the most important families which are over-represented in the Mabura Hill area (see table 1) and those families given by Huber (1995), Prance (1982), and Mori (1991) as characteristic Guianan lowland forest families. The origin of many families is given by Raven & Axelrod (1974), Axelrod & Raven (1978), and Gentry (1993), and listed in table 4a and 4b. Almost all families concerned have a West Gondwana-land origin, and only one, Malpighiaceae, diversified after the separation of South America and Africa. No truly Laurasian families are encountered within this group of families, although the origin of the Menispermaceae is questionable. For the Loganiaceae, the origin for the genus *Strychnos* is given; this is the genus found in our flora area, but the origin and taxonomic status of the family as a whole is unclear. These families all

Table 2.4
Origin of families that a) occur significantly more in the Mabura Hill area, and b) are cited as specific Guianan lowland forest components. Origin after Raven & Axelrod (1974), Axelrod & Raven (1978), Gentry (1993). Over-represented families derived from table 1, and specific Guianan lowlands families derived from Huber (1995), Prance (1991), and Mori (1991)

A: Over-represented families

Family	Origin		
	West Gwondanaland	Laurasia	South America
Annonaceae	x		
Apocynaceae	x		x
Bignoniaceae	x		
Chrysobalanaceae	x		
Connaraceae	x		
Cucurbitaceae	x		
Dilleniaceae	x		
Guttiferae	x		
Hippocrateaceae	x		
Lecythidaceae			x
Loganiaceae	x		
Malpighiaceae			x
Menispermaceae	x?	x	
Myristicaceae	x		
Nyctaginaceae			x
Sapindaceae	x		
Sapotaceae	x		
Simaroubaceae	x?		
Smilacaceae	??	??	??
Violaceae	x		x

B: Guianan lowland forest families

Family	Origin
	West Gwondanaland
Annonaceae	x
Burseraceae	x
Chrysobalanaceae	x
Lauraceae	x
Leguminosae	x
Meliaceae	x
Moraceae	x
Myristicaceae	x
Sapindaceae	x
Sapotaceae	x
Vochysiaceae	x

represent lineages of pre-Pleistocene age.

The forests of West Africa are generally similar in floristic composition to Neotropical ones (Gentry 1993), with Leguminosae as most important woody family group. The Caesalpinoideae, however, are more abundant in African floras than in the Neotropical floras (Gentry 1993). Diversity of Caesalpinoideae for Africa is highest in West Africa (Breteler pers. comm.), where in Ivory Coast the ratio Caesalpinoideae/Leguminosae amounts to 39% (de Koning 1983) and Gabon has a ratio of 42% (Pellegrin 1948, updated with Breteler pers. comm.). Locally even higher ratios can occur (e.g. in Gabon: 49% at Makokou (Gentry 1993), and 72% in specific plots (Reitsma 1988); and in Liberia: 73% at the Yoma-Gola National forest (Voorhoeve 1964)).

Within the Guayana Shield and Northern Brazil sites, the ratio Caesalpinoideae/Leguminosae is higher compared with other Neotropical areas (San Carlos de Río Negro 49%, Araracuara 33%, Mabura Hill 30%, Saül 28%, St. Elie 28%, Nouragues 23%, (the Guianas 30%) and Brazilian sites (Bahia 26% (Lewis 1987), Reserva Ducke 27%). As seen in Africa, locally even higher ratios can occur (e.g. 60 - 70 % in dry evergreen forest on white sands or 60% in Mora forest (Fanshawe 1952). It seems that especially in the area where the final separation between both continents occurred, the ratio of Caesalpinoideae within the Leguminosae is comparable.

Rietkerk *et al.(1995)* use the present distribution of Caesalpinoideae, both in species numbers and abundance, for delineation of evergreen lowland forest refuges in Central Africa. Dispersability of at least part of the Caesalpinoideae species is estimated to be extremely low and Rietkerk *et al.* (1995) hypothesized that areas with high species richness and/or abundance of Caesalpinoideae, are areas from which the species migrated. Extrapolating this hypothesis with our findings, we postulate that especially the Guayana Shield, and Northern Brazil, harbor the centre of origin of Caesalpinoideae for South America.

Connaraceae, very much over-represented in the Mabura Hill area with 13 species, have the highest genera diversity in Africa. In total this family has 12 genera, of which 10 occur in Africa, six in Asia, and four in tropical South America. South America, however, has the highest species diversity with a total of 101 spp., while Africa has a total of 49 species (Jongkind & Lemmens 1989). In the Mabura Hill area and in Saül all four South American genera were found. In the Western Amazonian sites, if cited, only one genus occurs.

Although most species of lowland rain forest predate the Pleistocene (Fjeldså & Lovett 1997), the large proportion of Caesalpinoideae in the Guayana Shield and Northern Brazilian sites, and the high percentage of species in such families as (in declining order) Connaraceae, Menispermaceae, Nyctaginaceae, Annonaceae, and Apocynaceae (also as climbers) (table 1), support our hypothesis of a high similarity with the forest flora of Western Africa.

In contrast to other regions in South America, the Guayana Plateau is characterized by its environmental stability, geologically (Pitman *et al.* 1993) and in terms of present day-ecological pressures: with no hurricane events, nor large-scale windthrows, or large scale river dynamics (Hammond & Brown 1995). Hammond & Brown (1995) showed that the mean seed weight of 187 native woody species of the Mabura Hill area, both trees and lianas, was significantly larger than that found in Manu and BCI (Mabura: mean fresh mass: 7.52 ± 1.44 g; Manu mean fresh mass: 1.68 ± 0.44; BCI: mean fresh mass 0.92 ± 0.38 g), with an enormous percentage of large seeded species (seed mass > 5 g) for the Mabura Hill area. This is true for the Leguminosae (mean fresh mass: 13.42 ± 4.64 g), and especially the Caesalpinoideae (mean fresh mass: 27.81 ± 12.66 g, with maximum up to 120 g). Large seeds are mainly mammal-, gravity and water-dispersed (Hammond & Brown 1995), and therefore supposed to have extremely low dispersability (sensu Rietkerk *et al.* 1995).

For the Mabura Hill area there are no indications of substantial historical Amerindian settlements (Evans & Meggers 1960). The French Guianan sites, however, have a flora that suggests historical Amerindian habitation. Furthermore, Saül has a long history of gold mining in the area. On a scale of disturbance by inhabitance Nouragues would be placed between Mabura Hill and Saül. The latter site has the lowest proportion of large-seeded families within the over-represented tree families; 14, 23, 28% for Saül, Nouragues and Mabura Hill, respectively. This again, suggests a correlation between the presence of large-seeded species and disturbance.

Conclusions

Compared with other sites the Mabura Hill area is not particularly rich in species, but the composition of the flora of Mabura Hill suggests that it is an area with a history of low natural disturbance. Only three introduced 'weedy' species were found. Consequently, the flora of Mabura Hill can still be considered an undisturbed Guianan lowland forest flora.

The flora of Mabura Hill, however, is also particular within Guyana, and within the Guianas. The growth form spectrum suggests that especially the species diversities of the herbs, shrubs, and epiphytes are relatively low, compared to other sites in the Guianas. The origin of the most important families in the Mabura Hill area, the large proportion of Caesalpinoideae in the Guayana Shield and Northern Brazil sites, and the high percentage of such families as (in declining order) Connaraceae, Menispermaceae, Nyctaginaceae, Annonaceae, and Apocynaceae (also as climbers) may still show traces of a common origin of the floras of (especially Northern) South America and Western Africa. Our data supports the hypothesis that especially the flora of the Guianas is strongly linked to the forest flora of Western Africa.

Acknowledgements

We would like to thank Demerara Timbers Ltd. for their hospitality and cooperation during our stay in Mabura Hill; H. Clark, J.A. Ratter, E. NicLughadha, O. Poncy, S.A. Mori, and G. Cremers for kindly supplying (unpublished) species lists; M. Polak and M.J. Jansen-Jacobs for compiling an earlier version of the checklist; E.A. Mennega for checking the species list, and M.J.A. Werger, P. Hogeweg, P.J.M. Maas, and R. Zagt for comments on an early version of this manuscript. During our stay in the field D. Hammond and S. Roberts kindly supplied us with flowering and fruiting specimens encountered and they contributed significantly to the assemblage of this list. This study was carried out within the framework of the Tropenbos-Guyana Programme, and funded by the Tropenbos Foundation, Wageningen, the Netherlands, Utrecht University, and the European Union.

Appendix 2.1.

Forest types represented in the inventory of the Mabura Hill area

Nr	name	drainage	location
	On Brown Sands:		
1	mixed forest / evergreen rain forest	poor	creek heads and valleys
2	mixed forest / evergreen rain forest	good	mid slope/ top slope
3	creek forest / evergreen rain forest	good	along small creeks
4	riverine floodplain forest / evergreen rain forest	partly inundated	broader flood plains
5	seasonal forest / semi-evergreenforest	good	gently undulating terrain
	On White Sands:		
6	dry evergreen forest / high, well-developed heath forest	excessive	
			fringing lower, swampy areas
7	dry evergreen low forest / low heath forest	excessive	
8	palm swamp forest / fresh water swamp forest	inundated	peaty soils
9	dry evergreen shrub / low, poorly developed heath forest	excessive	top watershed
	On laterite:		
10	mixed forest / evergreen rain forest good	excessive /	gravelly clay, laterite
11	xeric mixed forest	good	rocky soils

Appendix 2.1. (continued)

Nr	important species local name	scientific name	type
	On Brown Sands:		
1	Watafa	*Eperua rubiginosa*	dom
	Black kakaralli	*Eschweilera sagotiana*	co-dom
	Greenheart	*Chlorocardium rodiei*	co-dom
	Morabukea	*Mora gonggrijpii*	co-dom
2	Greenheart	*Chlorocardium rodiei*	dom \ co-dom
	Black kakaralli	*Eschweilera sagotiana*	dom \ co-dom
		Lecythidaceae spp.	dom \ co-dom
	Clump Wallaba	*Dicymbe altsonii*	dom \ co-dom
3	Soft Wallaba	*Eperua falcata*	abund
	Swamp Baromalli	*Catostemma commune*	abund
	Watafa	*Eperua rubiginosa*	abund
	Imirimiaballi	*Chamaecrista adiantifolia*	abund
	Barabara	*Diospyros ierensis*	abund
4	Mora	*Mora excelsa*	dom
	Watafa	*Eperua rubiginosa*	abund
	Crabwood	*Carapa guianensis*	abund
	Trysil	*Penthaclethra macroloba*	abund
	Corkwood	*Pterocarpus officinalis*	abund
	Sarebebe	*Macrolobium bifolium*	abund
5	Kabukalli	*Goupia glabra*	abund
	Simarupa	*Simarouba amara*	abund
	Koraroballi	*Hymenolobium* sp.	abund
	Iteballi	*Vochysia* spp.	abund
	Tatabu	*Diplotropis purpurea*	abund
	Leguminosae spp. (esp. Mimos.)	abund	
	Vochysiaceae spp.	abund	
	On White Sands:		
6	Soft Wallaba	*Eperua falcata*	dom
	Ituri Wallaba	*Eperua grandiflora*	dom
	Awasokule	*Tovomita* spp.	abund
	Itikiboroballi	*Swartzia* spp.	abund
	Yarula	*Aspidosperma excelsum*	abund
	Sand Baromalli	*Catostemma fragrans*	abund
	Korokororo / Horse eye	*Ormosia coutinhoi*	abund
7	Dakama	*Dimorphandra conjugata*	dom
	Duka	*Tapirira obtusa*	abund
8	Soft Wallaba	*Eperua falcata*	dom
	Ituri Wallaba	*Eperua grandiflora*	dom
	Charcoal Barabara	*Diospyros ierensis*	abund
	Turu	*Jessenia bataua*	abund
	White Cedar	*Tabebuia insignis*	abund
	Kirikaua	*Iryanthera lancifolia*	abund
	Manni	*Symphonia globulifera*	abund
	Wadara	*Couratari gloriosa*	abund
9	Dakama	*Dimorphandra conjugata*	dom
	Muri	*Humiria balsamifera* var. *guianensis*	dom
	On laterite:		
10	Morabukea	*Mora gonggrijpii*	abund
	Greenheart	*Chlorocardium rodiei*	abund
	Kakaralli	*Lecythis* or *Eschweilera* spp.	abund
	Itikiboroballi / Parakusan	*Swartzia* spp.	abund
	Sarebebeballi	*Vouacapoua macropetala*	abund
	Note: Unknown if species composition differs significantly from (1) or (2).		
11	Wild guavas	*Myrtaceae* spp.	dom
	Note: Further species composition unknown.		

Appendix 2

Checklist of plant of the vicinity of Mabura hill, Guyana, updated until June 1996.
compiled by: Renske C. Ek, Hans Ter Steege, Marion J. Jansen-Jacobs & Marcel Polak.

ACANTHACEAE
Beloperone violacea Planch. & Linden
Mendoncia squamuligera Nees
Odontonema schomburgkianum (Nees) Kuntze
Ruellia rubra Aubl.
Teliostachya alopecuroidea (Vahl) Nees

ALISMATACEAE
Echinodorus subalatus (Mart.) Griseb. subsp. ?

AMARANTHACEAE
Cyathula prostrata (L.) Blume

AMARYLLIDACEAE
Crinum erubescens L.f. ex Solander
Hippeastrum puniceum (Lam.) Baker

ANACARDIACEAE
Anacardium occidentale L.
Anacardium giganteum Hancock ex Engl.
Loxopterygium sagotii Hook.f.
Tapirira guianensis Aubl.
Tapirira obtusa (Benth.) J.D. Mitchell

ANNONACEAE
Anaxagorea dolichocarpa Sprague & Sandw.
Annona haematantha Miq.
Annona sericea Dunal
Annona symphyocarpa Sandw.
Annona sp.
Bocageopsis multiflora (Mart.) R.E. Fr.
Cymbopetalum brasiliense (Vell.) Benth. ex Baill.
Duguetia cadaverica Huber
Duguetia calycina Benoist
Duguetia inconspicua Sagot
Duguetia neglecta Sandw.
Duguetia pycnastera Sandw.
Duguetia yeshidan Sandw.
Duguetia sp.
Ephedranthus guianensis R.E. Fr.
Fusaea longifolia (Aubl.) Saff.
Guatteria atra Sandw.
Guatteria punctata (Aubl.) R.A. Howard
Guatteria schomburgkiana Mart.
Guatteria sp.
Oxandra asbeckii (Pulle) R.E. Fr.
Oxandra guianensis R.E. Fr.
Oxandra sp.
Rollinia exsucca (DC. ex Dunal) A. DC.
Trigynaea caudata (R.E. Fr.) R.E. Fr.
Trigynaea duckei (R.E. Fr.) R.E. Fr.
Unonopsis glaucopetala R.E. Fr.
Unonopsis guatterioides (A. DC.) R.E. Fr.
Unonopsis perrottetii (A. DC.) R.E. Fr.
Unonopsis rufescens (Baill.) R.E. Fr.
Unonopsis stipitata Diels
Xylopia aromatica (Lam.) Mart.
Xylopia benthamii R.E. Fr.
Xylopia nitida Dunal
Xylopia pulcherrima Sandw.
Xylopia sp.

APOCYNACEAE
Allamanda cathartica L.
Ambelania acida Aubl.
Aspidosperma album (Vahl) Benoist ex Pichon
Aspidosperma cruentum Woodson
Aspidosperma excelsum Benth.
Aspidosperma sandwithianum Markgr.
Aspidosperma vargasii A. DC.
Aspidosperma sp.
Forsteronia acouci (Aubl.) A. DC.
Forsteronia gracilis (Benth.) Muell. Arg.
Forsteronia guyanensis Muell. Arg.
Forsteronia schomburgkii A. DC. cf.
Geissospermum sericeum (Sagot) Benth. & Hook.f.
Himatanthus articulatus (Vahl) Woodson
Himatanthus articulatus x bracteatus
Himatanthus sucuuba (Spruce ex Muell. Arg.) Woodson
Lacmellea utilis (Arnott) Markgr.
Macoubea guianensis Aubl. var. guianensis
Malouetia gracilis (Benth.) A. DC.
Malouetia tamaquarina (Aubl.) A. DC.
Odontadenia amazonica (Stadelm.) Malme cf.
Odontadenia geminata (Roem. & Schult.) Muell. Arg.
Odontadenia perrottetii (A. DC.) Woodson
Odontadenia puncticulosa (A. Rich.) Pulle
Parahancornia fasciculata (Poir.) Benoist ex Pichon
Prestonia surinamensis Muell. Arg.
Tabernaemontana cerea (Woodson) Leeuwenberg
Tabernaemontana disticha DC.

Tabernaemontana meyeri (G.Mey. ex G. Don) Miers
Tabernaemontana undulata Vahl

AQUIFOLIACEAE
Ilex martiniana D. Don
Ilex sp.

ARACEAE
Anthurium gracile (Rudge) Lindl.
Anthurium hookeri Kunth
Anthurium jenmanii Engl.
Anthurium scolopendrinum (Ham.) Kunth
Anthurium trinerve Miq.
Anthurium sp.
Dieffenbachia paludicola N.E. Brown ex Gleason
Heteropsis flexuosa (Kunth) Bunting
Monstera dubia (Kunth.) Engl. & Krause
Monstera spruceana (Schott) Engl.
Montrichardia arborescens (L.) Schott
Philodendron callosum Krause
Philodendron ecordatum Schott
Philodendron fragrantissimum (Hook.) Kunth
Philodendron insigne Schott
Philodendron linnaei Kunth
Philodendron pedatum (Hook.) Kunth
Philodendron rudgeanum Schott
Philodendron scandens C. Koch & Sello
Philodendron solimoesense A.C. Sm.
Philodendron surinamense (Schott) Engl.
Philodendron sp.
Rhodospatha venosa Gleason
Spathiphyllum cuspidatum Schott
Spathiphyllum sp.
Stenospermation maguirei Jonker & A.M.E. Jonker
Urospatha sagittifolia (Rudge) Schott var. ?

ARALIACEAE
Dendropanax sp.
Schefflera decaphylla (Seemann) Harms

ARISTOLOCHIACEAE
Aristolochia consimilis Mast.
Aristolochia daemoninoxi Mast.
Aristolochia rugosa Lam.
Aristolochia weddellii Duch. var. rondoniana Hoehne

ASCLEPIADACEAE
Blepharodon nitidus (Vell.) Macbr.
Matelea palustris Aubl.
Matelea stenopetala Sandw.
Tassadia propinqua Decne.

BEGONIACEAE
Begonia humilis Dryand.

BIGNONIACEAE
Anemopaegma foetidum Bureau & K. Schum.
Anemopaegma oligoneuron (Sprague & Sandw.) A. Gentry
Anemopaegma parkeri Sprague
Arrabidaea candicans (Rich.) DC.
Arrabidaea egensis Bureau & K. Schum.
Arrabidaea fanshawei Sandw. cf.
Arrabidaea mollis (Vahl) Bureau ex K. Schum.
Arrabidaea patellifera (Schltdl.) Sandw.
Arrabidaea trailii Sprague
Callichlamys latifolia (Rich.) K. Schum.
Clytostoma binatum (Thunb.) Sandw.
Clytostoma sciuripabulum Bureau & K. Schum.
Cydista aequinoctialis (L.) Miers
Cydista sp.
Distictella elongata (Vahl) Urb.
Distictella magnoliifolia (Kunth) Sandw.
Distictella parkeri (DC.) Sprague & Sandw.
Jacaranda copaia (Aubl.) D. Don subsp. copaia
Lundia erionema DC.
Martinella obovata (Kunth) Bureau & K. Schum.
Memora bracteosa (DC.) Bureau & K. Schum. cf.
Memora flavida (DC.) Bureau & K. Schum.
Memora moringifolia (Miq.) Pulle
Memora racemosa A. Gentry
Memora schomburgkii (DC.) Miers
Memora sp.
Mussatia prieuriei (DC.) Bureau ex K. Schum.
Parabignonia steyermarkii Sandw.
Paragonia pyramidata (Rich.) Bureau
Phryganocydia corymbosa (Vent.) Bureau ex K. Schum.
Pleonotoma albiflora (Salzm. ex DC.) A. Gentry
Pleonotoma clematis (Kunth) Miers
Pleonotoma echitidea Sprague & Sandw.
Pleonotoma sp.
Potamoganos microcalyx (G. Mey.) Sandw.
Roentgenia sordida (Bureau & K. Schum.) Sprague & Sandw.
Schlegelia paraensis Ducke
Schlegelia violacea (Aubl.) Griseb.
Tabebuia insignis (Miq.) Sandw. var. monophylla Sandw.
Tabebuia serratifolia (Vahl) Nichols.

BOMBACACEAE
Catostemma commune Sandw.
Catostemma fragrans Benth.

Rhodognaphalopsis flaviflora (Pulle) A. Robijns var. flaviflora

BORAGINACEAE
Cordia bicolor A. DC.
Cordia exaltata Lam. var. melanoneura I.M. Johnston
Cordia fallax I.M. Johnston
Cordia nervosa Lam.
Cordia nodosa Lam.
Cordia sagotii I.M. Johnston
Cordia schomburgkii A. DC.
Cordia sp.
Heliotropium filiforme Lehm.
Tournefortia ulei Vaupel

BROMELIACEAE
Aechmea bromeliifolia (Rudge) Baker
Aechmea melinonii Hook.
Aechmea mertensii (Mey.) Schult.f.
Aechmea tillandsoides (Mart. ex Schult.f.) Baker
Ananas sp.
Bromelia plumieri (Griseb.) Mez
Catopsis berteroniana (Schult.f.) Mez
Guzmania lingulata (L.) Mez
Guzmania roezlii (E. Morren) Mez
Pitcairnia caricifolia Mart. ex Schult.f.
Streptocalyx longifolius (Rudge) Baker
Tillandsia anceps Lodd.
Tillandsia bulbosa Hook.
Tillandsia fasciculata Sw.
Tillandsia monadelpha (E. Morren) Baker
Tillandsia spiculosa Griseb. var. spiculosa
Tillandsia tenuifolia L
Tillandsia sp.
Vriesea amazonica (Baker) Mez
Vriesea gladioliflora (Wendl.) Antoine
Vriesea rubra (Ruiz & Pav.) Beer

BURMANNIACEAE
Campylosiphon purpurascens Benth.
Dictyostega orobanchoides (Hook.) Miers subsp. parviflora (Benth.) Snelders & Maas
Gymnosiphon breviflorus Gleason
Gymnosiphon divaricatus (Benth.) Benth. & Hook.f.
Gymnosiphon guianensis Gleason
Hexapterella gentianoides Urb.

BURSERACEAE
Protium aracouchini (Aubl.) Marchand
Protium decandrum (Aubl.) Marchand
Protium demerarense Swart
Protium guianense (Aubl.) Marchand var. guianense
Protium heptaphyllum (Aubl.) Marchand subsp. ?
Protium tenuifolium (Engl.) Engl.
Protium sp.
Tetragastris altissima (Aubl.) Swart
Trattinickia burserifolia Mart.
Trattinickia demerarae Sandw.
Trattinickia rhoifolia Willd.
Trattinickia sp.

CACTACEAE
Epiphyllum sp.
Rhipsalis sp.
Selenicereus sp.

CAPPARACEAE
Capparis maroniensis Benoist

CARYOCARACEAE
Caryocar nuciferum L.

CECROPIACEAE
Cecropia angulata I.W. Bailey
Cecropia obtusa Trécul.
Cecropia sciadophylla Mart.
Coussapoa latifolia Aubl.
Coussapoa microcephala Trécul.
Pourouma guianensis Aubl. subsp. guianensis

CELASTRACEAE
Goupia glabra Aubl.
Maytenus myrsinoides Reissek

CHRYSOBALANACEAE
Couepia bracteosa Benth.
Couepia cognata (Steud.) Fritsch
Couepia comosa Benth.
Couepia exflexa Fanshawe & Maguire
Couepia parillo DC.
Couepia sp.
Hirtella davisii Sandw.
Hirtella racemosa Lam. var. hexandra (Willd. ex Roem. & Schult.) Prance
Hirtella silicea Griseb.
Licania alba (Bernoulli) Cuatrec.
Licania boyanii Tutin
Licania buxifolia Sandw.
Licania canescens Benoist
Licania caudata Prance
Licania coriacea Benth.
Licania cuprea Sandw.
Licania densiflora Kleinh.
Licania divaricata Benth.
Licania fanshawei Prance
Licania guianensis (Aubl.) Griseb.
Licania heteromorpha Benth. var. heteromorpha
Licania heteromorpha Benth. var. perplexans Sandw.
Licania hypoleuca Benth.
Licania incana Aubl.
Licania intrapetiolaris Spruce ex Hook.f.
Licania laevigata Prance
Licania laxiflora Fritsch
Licania leptostachya Benth.
Licania majuscula Sagot
Licania membranacea Sagot ex Laness.
Licania micrantha Miq.
Licania persaudii Fanshawe & Maguire
Licania piresii Prance
Licania sp.

Parinari campestris Aubl.
Parinari excelsa Sabine
Parinari parvifolia Sandw.
Parinari rodolphii Huber
Parinari sp.

COMBRETACEAE
Buchenavia fanshawei Exell & Maguire
Buchenavia grandis Ducke
Combretum laxum Jacq.
Combretum rotundifolium Rich.
Terminalia amazonica (J.F. Gmel.) Exell
Terminalia dichotoma G. Mey.
Terminalia cf. guyanensis Eichler

COMMELINACEAE
Commelina erecta L.
Commelina virginica L.

COMPOSITAE
Acanthospermum australe (Loefl.) Kuntze
Bidens cynapiifolia Kunth
Calea caleoides (DC.) H. Rob.
Chromolaena odorata (L.) R. King & H. Rob.
Clibadium sylvestre (Aubl.) Baill.
Conyza bonariensis (L.) Cronquist
Cyanthillium cinerea (L.) H. Rob.
Emilia coccinea (Sims) G. Don
Emilia sonchifolia (L.) DC. ex Wight
Erechtites hieracifolia (L.) Raf. ex DC.
Lepidaploa gracilis (Kunth) H. Rob.
Lepidaploa remotiflora (Rich.) H. Rob.
Mikania gleasonii B.L. Rob.
Mikania hookeriana DC.
Mikania psilostachya DC.
Porophyllum ruderale (Jacq.) Cass.
Unxia camphorata L.f.
Wulffia baccata (L.f.) Kuntze

CONNARACEAE
Cnestidium guianense (Schellenb.) Schellenb.
Connarus coriaceus Schellenb.
Connarus erianthus Benth. ex Baker var. stipitatus Forero
Connarus megacarpus S.F. Blake
Connarus perrottetii (DC.) Planch. var. perrottetii
Connarus perrottetii (DC.) Planch. var. rufus Forero
Connarus punctatus Planch.
Connarus sp.
Pseudoconnarus macrophyllus (Poepp. & Endl.) Radlk.
Rourea frutescens Aubl.
Rourea induta Planch. var. induta
Rourea ligulata Baker cf.
Rourea pubescens (DC.) Radlk. var. spadicea (Radlk.) Forero

CONVOLVULACEAE
Dicranostyles guianensis Mennega
Ipomoea philomega (Vell.) House
Ipomoea tiliacea (Willd.) Choisy
Jacquemontia tamnifolia (L.) Griseb.
Lysiostyles scandens Benth.
Maripa glabra Choisy cf.
Maripa scandens Aubl.
Merremia umbellata (L.) H. Hallier

CUCURBITACEAE
Cayaponia ophthalmica R.E. Schultes
Cayaponia racemosa (Mill.) Cogn.
Cayaponia selysioides C. Jeffrey
Citrullus lanatus (Thunb.) Matsumura & Nakai
Cucurbita moschata (Duchesne ex Lam.) Duchesne ex Poir.
Gurania bignoniacea (Poepp. & Endl.) C. Jeffrey
Gurania spinulosa (Poepp. & Endl.) Cogn.
Gurania subumbellata (Miq.) Cogn.
Gurania sp.
Helmontia leptantha (Schltd.) Cogn.
Lagenaria siceraria (Molina) Standl.
Momordica charantia L.
Psiguria triphylla (Miq.) C. Jeffrey
Psiguria sp.

CYCLANTHACEAE
Cyclanthus bipartitus Poit.
Evodianthus funifer (Poit.) Lindm. subsp. funifer
Evodianthus funifer (Poit.) Lindm. subsp. trailanus (Drude) Harling
Thoracocarpus bissectus (Vell.) Harling

CYPERACEAE
Becquerelia cymosa Brongn. subsp. ?
Calyptrocarya glomerulata (Brogn.) Urb. var. glomerulata
Cyperus ligularis L.
Cyperus luzulae (L.) Retz.
Cyperus simplex Kunth
Cyperus sphacelatus Rottb.
Cyperus surinamensis Rottb.
Diplasia karataefolia Rich.
Fimbristylis annua (All.) Roem. & Schult.
Fimbristylis littoralis Gaudich.
Fimbristylis miliacea (L.) Vahl
Hypolytrum amplum Poepp. & Endl. ex Kunth
Hypolytrum jenmanii C.B. Clarke
Kyllinga odorata Vahl
Kyllinga sp.
Lagenocarpus verticillatus (Spreng.) T. Koyama & Maguire
Mapania sylvatica Aubl. subsp. sylvatica
Pycreus polystachyos (Rottb.) P. Beauv.
Rhynchospora cephalotes (L.) Vahl

Rhynchospora holoschoenoides (Rich.) Herter
Rhynchospora pubera (Vahl) Boeck.
Rhynchospora watsonii (Britt.) Davidse
Scleria pterota Presl
Scleria stipularis Nees
Scleria sp.

DICHAPETALACEAE
Dichapetalum pedunculatum (DC.) Baill.
Dichapetalum rugosum (Vahl) Prance
Tapura capitulifera Baill.
Tapura guianensis Aubl.

DILLENIACEAE
Davilla kunthii St.Hil.
Doliocarpus brevipedicellatus Garcke subsp. brevipedicellatus
Doliocarpus guianensis (Aubl.) Gilg
Doliocarpus macrocarpus Mart. ex Eichler
Doliocarpus major J.F. Gmel. subsp. major
Doliocarpus paraensis Sleumer
Doliocarpus spraguei Cheesem.
Pinzona coriacea Mart.& Zucc.
Tetracera asperula Miq. subsp. asperula
Tetracera tigarea DC.
Tetracera volubilis L. subsp. volubilis

DIOSCOREACEAE
Dioscorea amazonicum Mart. ex Griseb.
Dioscorea crotalariifolia Uline
Dioscorea dodecaneura Vell.
Dioscorea megacarpa Gleason
Dioscorea pilosiuscula Bert. ex Spreng.
Dioscorea samydea Mart. ex Griseb.
Dioscorea sp.

EBENACEAE
Diospyros dichroa Sandw.
Diospyros guianensis (Aubl.) Gürke subsp. guianensis
Diospyros ierensis Britton
Diospyros virginiana L.
Diospyros sp.

ELAEOCARPACEAE
Sloanea eichleri K. Schum. cf.
Sloanea grandiflora Sm.
Sloanea guianensis (Aubl.) Benth.
Sloanea laurifolia (Benth.) Benth. aff.
Sloanea sp.

ERIOCAULACEAE
Eriocaulon heterodoxum Moldenke
Paepalanthus bifidus (Schrad.) Kunth
Paepalanthus lamarckii Kunth

ERYTHROXYLACEAE
Erythroxylum citrifolium St. Hil.
Erythroxylum macrophyllum Cav.
Erythroxylum squamatum Sw.

EUPHORBIACEAE
Alchornea triplinervia (Spreng.) Muell. Arg.
Alchorneopsis floribunda (Benth.) Muell. Arg.
Amanoa guianensis Aubl.
Aparisthmium cordatum (A. Juss.) Baill.
Chaetocarpus schomburgkianus (Kuntze) Pax & K. Hoffm.
Chamaesyce hirta (L.) Millsp.
Chamaesyce serpens (Kunth) Small var. serpens
Chamaesyce thymifolia (L.) Millsp.
Conceveiba guianensis Aubl.
Croton trinitatis Millsp.
Croton sp. nov.?
Croton sp.
Dalechampia olympiana Kuhlm. & Rodr.
Dalechampia parvibracteata Lanj.
Glycydendron amazonicum Ducke
Hevea pauciflora (Spruce ex Benth.) Muell.Arg. var. coriacea Ducke
Hyeronima alchorneoides Allemão var. alchorneoides
Hyeronima oblonga (Tul.) Muell. Arg.
Mabea speciosa Muell. Arg. subsp. concolor (Muell. Arg.) Den Hollander
Mabea speciosa Muell. Arg. subsp. guianensis Esser
Mabea sp.
Maprounea guianensis Aubl.
Micrandra elata (Didr.) Muell. Arg.
Pera bicolor (Klotzsch) Muell. Arg.
Pera sp.
Phyllanthus stipulatus (Raf.) Webster
Plukenetia volubilis L.
Sandwithia guyanensis Lanj.
Sandwithia lessertiana (Baill.) Muell. Arg.
Sebastiana corniculata (Vahl) Muell. Arg.

FLACOURTIACEAE
Carpotroche surinamensis Uittien
Casearia combaymensis Tul.
Casearia commersoniana Camb.
Casearia javitensis Kunth
Casearia pitumba Sleumer
Casearia sp.
Homalium guianense (Aubl.) Oken
Homalium racemosum Jacq.
Laetia procera (Poepp.) Eichler
Ryania pyrifera (Rich.) Sleumer & Uittien
Ryania speciosa Vahl var. tomentosa (Miq.) Monach.

GENTIANACEAE
Coutoubea ramosa Aubl. var. ramosa
Irlbachia alata (Aubl.) Maas subsp. alata

Irlbachia purpurascens (Aubl.) Maas
Voyria aphylla (Jacq.) Pers.
Voyria caerulea Aubl.
Voyria clavata Splitg.
Voyria corymbosa Splitg. subsp. alba (Standley) Ruyters & Maas
Voyria corymbosa Splitg. subsp. corymbosa
Voyria rosea Aubl.
Voyria tenuiflora Griseb. cf.
Voyria sp.
Voyriella parviflora (Miq.) Miq.

GESNERIACEAE
Besleria laxiflora Benth.
Besleria verecunda C.V. Morton
Chrysothemis pulchella (Donn ex Sims) Decne.
Codonanthe calcarata (Miq.) Hanst.
Codonanthe crassifolia (Focke) C.V. Morton
Drymonia serrulata (Jacq.) Mart.
Nautilocalyx pictus (Hook.) Sprague
Paradrymonia densa (C.H. Wright) Wiehler
Paradrymonia maculata (Hook.f.) Wiehler

GNETACEAE
Gnetum nodiflorum Brongn.
Gnetum urens (Aubl.) Blume
Gnetum sp.

GRAMINEAE
Andropogon bicornis L.
Andropogon leuchostachyus Kunth
Andropogon selloanus (Hack.) Hack.
Axonopus capillaris (Lam.) Chase
Cenchrus brownii Roem. & Schult.
Cenchrus echinatus L.
Digitaria lanuginosa (Nees) Henr.
Eragrostis maypurensis (Kunth) Steud.
Gymnopogon foliosus (Willd.) Nees
Gynerium sagittatum (Aubl.) P. Beauv.
Homolepis isocalycia (G. Mey.) Chase
Ichnanthus nemoralis (Schrad.) Hitchc. & Chase
Ichnanthus pallens (Sw.) Munro ex Benth.
Lasiacis ligulata Hitchc. & Chase
Olyra micrantha Kunth
Orthoclada laxa (Rich.) P. Beauv.
Panicum pilosum Sw.
Panicum stoloniferum Poir.
Paspalum decumbens Sw.
Paspalum melanospermum Desv. ex Poir.

GUTTIFERAE
Calophyllum brasiliense Cambess.
Caraipa sp.
Clusia cuneata Benth.
Clusia fockeana Miq.
Clusia grandiflora Splitg.
Clusia jenmanii Engl.
Clusia myriandra (Benth.) Planch. & Triana
Clusia nemorosa G. Mey.
Clusia palmicida Rich.
Clusia pana-panari (Aubl.) Choisy
Clusia schomburgkiana (Planch. & Triana) Benth. ex Engl.
Clusia sp.
Clusiella elegans Planch. & Triana
Havetiopsis flavida (Benth.) Planch. & Triana
Platonia insignis Mart.
Quapoya bracteolata Sandw.
Quapoya peruviana (Poepp. & Endl.) O. Kuntze
Rheedia benthamiana Planch. & Triana
Symphonia globulifera L.f.
Tovomita brevistaminea Engl.
Tovomita calodictyos Sandw.
Tovomita cephalostigma Vesque
Tovomita tenuiflora Benth.
Tovomita sp.
Vismia guianensis (Aubl.) Choisy var. guianensis
Vismia macrophylla Kunth
Vismia sessilifolia (Aubl.) Choisy
Vismia sp.

HIPPOCRATEACEAE
Cheiloclinium cognatum (Miers) A.C. Sm.
Cheiloclinium diffusiflorum (Peyr.) A.C. Sm.
Cheiloclinium hippocrateoides (Peyr.) A.C. Sm.
Cheiloclinium sp.
Hippocratea volubilis L.
Peritassa compta Miers
Peritassa laevigata (Hoffmanns.) A.C. Sm.
Prionostemma aspera (Lam.) Miers
Salacia maburensis Mennega
Salacia multiflora (Lam.) DC. subsp. multiflora
Tontelea attenuata Miers
Tontelea coriacea A.C. Sm.
Tontelea laxiflora (Benth.) A.C. Sm.
Tontelea nectandrifolia (A.C. Sm.) A.C. Sm.
Tontelea ovalifolia (Miers) A.C. Sm.

HUMIRIACEAE
Humiria balsamifera (Aubl.) A. St.Hil. var. balsamifera
Humiria balsamifera (Aubl.) A. St.Hil. var. floribunda (Mart.) Cuatrec.
Humiria balsamifera (Aubl.) A. St.Hil. var. guianensis (Benth.) Cuatrec.
Humiriastrum obovatum (Benth.) Cuatrec.
Sacoglottis guianensis Benth. var. guianensis

ICACINACEAE
Discophora guianensis Miers

Emmotum fagifolium Desv.
Leretia cordata Vell.
Pleurisanthes flava Sandw.

LABIATAE
Hyptis atrorubens Poit.

LACISTEMACEAE
Lacistema aggregatum (Bergius) Rusby
Lacistema sp.

LAURACEAE
Aiouea guianensis Aubl.
Aniba canelilla (Kunth) Mez
Aniba citrifolia (Nees) Mez
Aniba excelsa Kosterm.
Aniba hypoglauca Sandw.
Aniba kappleri Mez
Aniba panurensis (Meissn.) Mez
Aniba rosaeodora Ducke cf.
Chlorocardium rodiei (Schomb.) Rohwer, Richter & v.d. Werff
Endlicheria chalisea Chanderballi
Endlicheria multiflora (Miq.) Mez
Endlicheria punctulata (Mez) C.K. Allen
Licaria cannella (Meissn.) Kosterm.
Nectandra amazonum Nees
Ocotea aciphylla (Nees) Mez
Ocotea acutangula (Miq.) Mez
Ocotea canaliculata (Rich.) Mez
Ocotea caudata (Nees) Mez aff.
Ocotea cymbarum Kunth
Ocotea discrepens C.K. Allen cf.
Ocotea floribunda (Sw.) Mez
Ocotea glomerata (Nees) Mez
Ocotea guianensis Aubl. var. guianensis
Ocotea pauciflora (Nees) Mez
Ocotea puberula (Rich.) Nees
Ocotea schomburgkiana (Nees) Mez
Ocotea tomentella Sandw.
Ocotea sp.
Ocotea sp. nov.

LECYTHIDACEAE
Couratari gloriosa Sandw.
Couratari guianensis Aubl.
Eschweilera alata A.C. Sm.
Eschweilera coriacea (A. DC.) Mori
Eschweilera decolorans Sandw.
Eschweilera micrantha (Berg) Miers
Eschweilera parviflora (Aubl.) Miers
Eschweilera pedicellata (Rich.) Mori
Eschweilera sagotiana Miers
Eschweilera subglandulosa (Steud. ex Berg) Miers
Eschweilera wachenheimii (Benoist) Sandw.
Eschweilera sp.
Gustavia augusta L.
Lecythis alutacea (A.C. Sm.) Mori
Lecythis confertiflora (A.C. Sm.) Mori
Lecythis corrugata Poit. subsp. corrugata
Lecythis holcogyne (Sandw.) Mori
Lecythis zabucajo Aubl.

LEGUMINOSAE (CAESALPINIODEAE)
Bauhinia guianensis Aubl. var. guianensis
Bauhinia scala-simiae Sandw.
Bauhinia siqueiraei Ducke
Bauhinia surinamensis Amsh.
Bauhinia sp.
Campsiandra comosa Benth. subsp. comosa
Cassia cowanii Irwin & Barneby var. guianensis (Sandw.) Irwin & Barneby
Chamaecrista adiantifolia (Benth.) Irwin & Barneby var. pteridophylla Irwin & Barneby
Chamaecrista apoucouita (Aubl.) Irwin & Barneby
Crudia glaberrima (Steud.) Macbr.
Dicymbe altsonii Sandw.
Dimorphandra conjugata (Splitg.) Sandw.
Dimorphandra polyandra Benoist
Elizabetha coccinea M.R. Schomb. ex Benth. var. oxyphylla (Harms) R.S. Cowan
Elizabetha durissima Ducke
Eperua falcata Aubl.
Eperua grandiflora (Aubl.) Benth. subsp. guyanensis R.S. Cowan
Eperua jenmanii Oliv. subsp. sandwithii R.S. Cowan
Eperua rubiginosa Miq.
Eperua schomburgkiana Benth.
Hymenaea courbaril L. var. courbaril
Macrolobium bifolium (Aubl.) Pers.
Macrolobium jenmanii (Gleason) Sandw.
Macrolobium montanum Ducke var. potaroanum R.S. Cowan
Mora excelsa Benth.
Mora gonggrijpii (Kleinh.) Sandw.
Paloue guianensis Aubl.
Paloue riparia Pulle
Peltogyne paniculata Benth. subsp. pubescens (Benth.) M.F. Silva
Peltogyne venosa (Vahl) Benth. subsp. densiflora (Spruce ex Benth.) M.F. Silva
Peltogyne venosa (Vahl) Benth. subsp. venosa
Sclerolobium guianense Benth. var. guianense
Senna latifolia (G. Mey.) Irwin & Barneby
Senna quinquangulata (L.C. Rich.) Irwin & Barneby
Tachigali pubiflora Benth.
Tachigali rusbyi Harms cf.
Tachigali sp.
Vouacapoua americana Aubl.
Vouacapoua macropetala Sandw.

LEGUMINOSAE (MIMOSOIDEAE)
Abarema jupunba (Willd.) Britton & Killip var. trapezifolia Barneby & Grimes
Acacia paraensis Ducke
Balizia pedicellaris (DC.) Barneby & Grimes
Enterolobium barnebianum Mesquita cf.
Enterolobium cyclocarpum (Jacq.) Griseb. cf.
Inga alba (Sw.) Willd.
Inga disticha Benth.
Inga huberi Ducke
Inga lateriflora Miq.
Inga meissneriana Miq.
Inga nobilis Willd.
Inga sertulifera DC.
Inga stipularis DC.
Inga thibaudiana DC.
Inga sp.
Macrosamanea discolor (Humb. & Bonpl.) Britton & Rose ex Britton & Killip
Macrosamanea pubiramea (Steud.) Barneby & Grimes
Mimosa myriadenia (Benth.) Benth. var. myriadenia
Mimosa sp.
Parkia nitida Miq.
Parkia pendula (Willd.) Benth. ex Walp.
Parkia ulei (Harms) Kuhlm. var. surinamensis Kleinh.
Pentaclethra macroloba (Willd.) Kuntze
Pseudopiptadenia suavolens (Miq.) Grimes
Zygia latifolia (L.) Fawc. & Rendle
Zygia racemosa (Ducke) Barneby & Grimes

LEGUMINOSAE (PAPILIONOIDEAE)
Acosmium nitens (Vogel) Yakovlev
Acosmium praeclarum (Sandw.) Yakovlev
Aeschynomene sensitiva Sw.
Alexa imperatricis (R.H. Schomb.) Baill.
Alexa leiopetala Sandw.
Alexa sp.
Andira coriacea Pulle
Calopogonium mucunoides Desv.
Centrosema latissimum Ducke
Centrosema triquetrum Spruce ex Benth.
Clathrotropis brachypetala (Tul.) Kleinh.
Clathrotropis macrocarpa Ducke
Clitoria leptostachya Benth.
Clitoria sagotii Fantz var. sagotii
Clitoria sagotii Fantz var. canaliculata Fantz
Clitoria sp.
Dalbergia ecastophyllum (L.) Taub.
Dalbergia riedelii (Radlk.) Sandw.
Desmodium adscendens (Sw.) DC.
Desmodium axillare (Sw.) DC.
Desmodium barbatum (L.) Benth.
Dioclea elliptica R.H. Maxwell
Dioclea scabra (Rich.) R.H. Maxwell var. scabra
Dioclea sp.
Diplotropis purpurea (Rich.) Amsh. var. purpurea
Dipteryx odorata (Aubl.) Willd.
Hymenolobium flavum Kleinh. cf.
Hymenolobium petraeum Ducke cf.
Hymenolobium sp. nov.
Lonchocarpus negrensis Benth.
Machaerium aureiflorum Ducke
Machaerium ferox (Mart. ex Benth.) Ducke
Machaerium madeirense Pittier
Machaerium multisii Killip ex Rudd
Machaerium myrianthum Spruce ex Benth.
Machaerium oblongifolium Vogel
Machaerium quinata (Aubl.) Sandw. var. parviflorum (Benth.) Rudd
Machaerium quinata (Aubl.) Sandw. var. quinata
Machaerium sp.
Ormosia amazonica cf.
Ormosia coarctata Jacks.
Ormosia coccinea (Aubl.) Jacks.
Ormosia costulata (Miq.) Kleinh.
Ormosia coutinhoi Ducke
Ormosia paraensis Ducke
Poecilanthe hostmannii (Benth.) Amsh.
Pterocarpus rohrii Vahl
Pterocarpus santalinoides L'Hér. ex DC.
Swartzia benthamiana Miq. var. benthamiana
Swartzia grandifolia Bong. ex Benth.
Swartzia guianensis (Aubl.) Urb.
Swartzia jenmanii Sandw.
Swartzia laevicarpa Amsh.
Swartzia leiocalycina Benth.
Swartzia leiogyne (Sandw.) R.S. Cowan
Swartzia longipedicellata Sandw.
Swartzia oblanceolata Sandw.
Swartzia roraimae Sandw.
Swartzia schomburgkii Benth. var. schomburgkii
Swartzia sprucei Benth. var. tessellata R.S. Cowan
Swartzia xanthopetala Sandw.
Swartzia sp.
Swartzia sp. nov.
Vatairea guianensis Aubl.
Vigna luteola (Jacq.) Benth.
Zornia latifolia Sm. var. latifolia

LENTIBULARIACEAE
Utricularia pusilla Vahl

LINACEAE
Hebepetalum humiriifolium (Planch.) Benth.
Roucheria schomburgkii Planch.

LISSOCARPACEAE
Lissocarpa guianensis Gleason
Lissocarpa sp.

LOGANIACEAE
Antonia ovata Pohl
Spigelia humilis Benth.
Spigelia multispica Steud.
Strychnos bredemeyeri (Schult.) Sprague & Sandw.
Strychnos diaboli Sandw.
Strychnos erichsonii M.R. Schomb.
Strychnos guianensis (Aubl.) Mart.
Strychnos hirsuta Spruce ex Benth.
Strychnos melinoniana Baill.
Strychnos mitscherlichii M.R. Schomb. var. mitscherlichii
Strychnos subcordata Spruce
Strychnos tomentosa Benth.
Strychnos sp.

LORANTHACEAE
Oryctanthus florulentus (Rich.) Urb.
Phthirusa disjectifolia (Rizzini) Kuijt
Phthirusa guyanensis Eichler
Phthirusa retroflexa (Ruiz. & Pav.) Kuijt
Phthirusa rufa (Mart.) Eichler
Phthirusa stelis (L.) Kuijt
Phthirusa sp.
Struthanthus sp.

LYCOPODIACEAE
Huperzia dichotoma (Jacq.) Trevis.
Huperzia sp.
Lycopodiella cernua (L.) Pic.-Serm.

MALPIGHIACEAE
Banisteriopsis martiniana (A. Juss.) Cuatrec. var. martiniana
Burdachia sphaerocarpa A. Juss.
Byrsonima aerugo Sagot
Byrsonima crassifolia (L.) Kunth
Byrsonima eugeniifolia Sandw.
Byrsonima gymnocalycina A. Juss.
Byrsonima stipulacea A. Juss.
Heteropterys cristata Benth. cf.
Heteropterys macradena (DC.) W.R. Anderson
Heteropterys macrostachya A. Juss.
Heteropterys multiflora (DC.) Hochr.
Heteropterys nervosa A. Juss.
Hiraea adenophora Sandw. cf.
Hiraea affinis Miq.
Hiraea faginea (Sw.) Nied.
Hiraea fagifolia (DC.) A. Juss.
Mascagnia guianensis W.R. Anderson
Mascagnia sepium (A. Juss.) Griseb.
Mezia includens (Benth.) Cuatrec.
Stigmaphyllon sinuatum (DC.) A. Juss.
Tetrapterys acutifolia Cav.
Tetrapterys calophylla A. Juss.
Tetrapterys crispa A. Juss.
Tetrapterys fimbripetala A. Juss.
Tetrapterys mucronata Cav.
Tetrapterys styloptera A. Juss.

MALVACEAE
Abutilon indicum (L.) Sweet
Hibiscus sabdariffa L.
Sida glomerata Cav.
Urena lobata L.

MARANTACEAE
Ischnosiphon foliosus Gleason
Ischnosiphon gracilis (Rudge) Körn.
Ischnosiphon obliquus (Rudge) Körn.
Ischnosiphon puberulus Loes. var. scaber (Petersen) L. Andersson
Monotagma spicatum (Aubl.) J.F. Macbr.

MARCGRAVIACEAE
Marcgravia coriacea Vahl
Marcgravia parviflora Rich. ex Wittm.
Marcgravia sp.
Norantea guianensis Aubl.
Norantea sp.
Souroubea guianensis Aubl. subsp. guianensis

MELASTOMATACEAE
Aciotis circaeifolia (Bonpl.) Triana
Aciotis fragilis (DC.) Cogn.
Aciotis laxa (DC.) Cogn. var. kappleriana (Naud.) Cogn.
Aciotis laxa (DC.) Cogn. var. laxa
Aciotis ornata (Miq.) Gleason
Aciotis purpurascens (Aubl.) Triana
Bellucia grossularioides (L.) Triana
Clidemia bullosa DC.
Clidemia conglomerata DC.
Clidemia involucrata DC.
Clidemia minutiflora (Triana) Cogn.
Clidemia saulensis Wurdack aff.
Clidemia sp.
Comolia vernicosa (Benth.) Triana
Henriettea multiflora Naudin
Henriettea ramiflora (Sw.) DC.
Henriettea stellaris O. Berg ex Triana
Leandra divaricata (Naudin) Cogn.
Leandra purpurea Gleason
Loreya mespiloides Miq.
Macrocentrum cristatum (DC.) Triana var. parviflorum (DC.) Cogn.
Miconia argyrophylla DC. subsp. argyrophylla
Miconia argyrophylla DC. subsp. gracilis Wurdack
Miconia bracteata (DC.) Triana
Miconia ceramicarpa (DC.) Cogn. var. ceramicarpa
Miconia gratissima Benth. ex Triana
Miconia hypoleuca (Benth.) Triana
Miconia mirabilis (Aubl.) L.O. Williams

Miconia nervosa (Sm.) Triana
Miconia prasina (Sw.) DC. aff.
Miconia punctata (Desr.) D. Don
Miconia rubiginosa (Bonpl.) DC.
Miconia serrulata (DC.) Naudin
Mouriri acutiflora Naudin
Mouriri sp.
Tibouchina aspera Aubl. var. asperrima Cogn.
Tococa acuminata Benth.
Tococa aristata Benth.
Tococa subciliata (DC.) Triana
Tococa sp.

MELIACEAE
Carapa guianensis Aubl.
Carapa procera A. DC.
Cedrela odorata L.
Guarea kunthiana A. Juss.
Guarea pubescens (Rich.) A. Juss. subsp. pubescens
Trichilia sp.

MENISPERMACEAE
Abuta barbata Miers
Abuta bullata Moldenke
Abuta candollei Triana & Planch.
Abuta grandifolia (Mart.) Sandw. cf.
Abuta imene Eichler
Abuta obovata Diels
Abuta rufescens Aubl.
Abuta sandwithiana Krukoff & Barneby
Abuta velutina Gleason
Abuta sp.
Anomospermum grandifolium Eichler
Anomospermum sp.
Cissampelos andromorpha DC.
Curarea candicans (Rich.) Barneby & Krukoff
Odontocarya tamoides (DC.) Miers
Sciadotenia cayennensis Benth.
Telitoxicum inopinatum Moldenke
Telitoxicum krukovii Moldenke
Telitoxicum minutiflorum (Diels) Moldenke
Telitoxicum sp
indet. sp.

MONIMIACEAE
Siparuna decipiens (Tul.) A. DC.
Siparuna guianensis Aubl.
Siparuna sp.

MORACEAE
Bagassa guianensis Aubl.
Brosimum guianense (Aubl.) Huber
Brosimum rubescens Taub.
Clarisia ilicifolia (Spreng.) Lanj. & Rossberg
Ficus gomelleira Kunth & Bouché
Ficus guianensis Desv.
Ficus obtusifolia Kunth
Ficus paludica Standl.
Ficus panurensis Standl.
Ficus paraensis (Miq.) Miq.
Ficus sp.
Helicostylis tomentosa (Poepp. & Endl.) Rusby
Naucleopsis guianensis (Mildbr.) C.C. Berg
Trymatococcus paraensis Ducke

MUSACEAE
Heliconia acuminata Rich.
Heliconia bihai (L.) L.
Heliconia chartacea Lane ex Barreiros
Heliconia densiflora Verlot subsp. angustifolia L. Andersson
Heliconia sp.

MYRISTICACEAE
Iryanthera lancifolia Ducke
Iryanthera sagotiana (Benth.) Warb.
Otoba sp.
Virola elongata (Benth.) Warb.
Virola michelii Heckel
Virola sebifera Aubl.
Virola surinamensis (Rol. ex Rottb.) Warb.

MYRSINACEAE
Cybianthus fulvopulverulenta (Mez) A. Agostini
Cybianthus gracillimus (Warm.) Mez
Cybianthus guianensis (A. DC.) Miq. subsp. pseudoicacoreus (A. DC.) A. Agostini
Cybianthus prieurii A. DC.
Cybianthus surinamensis (Spreng.f.) A. Agostini
Cybianthus venezuelanus Mez
Stylogyne schomburgkiana (DC.) Mez

MYRTACEAE
Calycolpus goetheanus (Mart. ex DC.) O. Berg
Calycolpus revolutus (Schauer) O. Berg
Calyptranthes forsteri O. Berg
Calyptranthes sp.
Eugenia coffeifolia DC. vel aff.
Eugenia conjuncta Amsh.
Eugenia cucullata Amsh. aff.
Eugenia excelsa O. Berg
Eugenia patrisii Vahl
Eugenia tapacumensis O. Berg
Eugenia trinervia Vahl
Eugenia sp.
Marlierea montana (Aubl.) Amsh.
Marlierea schomburgkiana O. Berg
Myrcia fallax (Rich.) DC.
Myrcia subobliqua (Benth.) Nied.
Myrcia sylvatica (G. Mey.) DC.
Myrcia sp.
Myrcia sp. [sect. Aulomyrcia]
Myrciaria floribunda (West ex Willd.) O. Berg
Myrciaria vismeifolia (Benth.) O. Berg
Psidium guajava L.
Psidium persoonii McVaugh
Psidium striatulum DC.

NYCTAGINACEAE
Guapira eggersiana (Heimerl) Lundell

Guapira fragrans (Dum.Cours.)Little
Neea constricta Spruce ex Schmidt
Neea floribunda Poepp. & Endl.
Neea sp.
Pisonia glabra Heimerl

OCHNACEAE
Cespedesia spathulata (Ruiz & Pav.) Planch.
Ouratea cardiosperma (Lam.) Engl. aff.
Ouratea guianensis Aubl.
Ouratea leblondii (Tiegh.) Lemée
Ouratea sp.
Sauvagesia erecta L. subsp. erecta

OLACACEAE
Chaunochiton kappleri (Sagot ex Engl.) Ducke
Dulacia guianensis (Engl.) Kuntze
Heisteria cauliflora Sm.
Maburea trinervis Maas
Minquartia guianensis Aubl.

ONAGRACEAE
Ludwigia decurrens Walter
Ludwigia erecta (L.) Hara
Ludwigia hyssopifolia (G. Don) Exell

OPILIACEAE
Agonandra silvatica Ducke

ORCHIDACEAE
Batemannia colleyi Lindl.
Bifrenaria longicornis Lindl.
Brassavola martiana Lindl.
Brassia verrucosa Lindl. (= B. bidens Lindl.?)
Brassia wageneri Rchb.f.
Campylocentrum fasciola (Lindl.) Cogn.
Campylocentrum micranthum (Lindl.) Rolfe
Catasetum barbatum (Lindl.) Lindl. var. barbatum
Catasetum discolor (Lindl.) Lindl.
Caularthron bicornutum (Hook.) Raf.
Cheiradenia imthurnii Cogn.
Coryanthes sp.
Dichaea muricata (Sw.) Lindl.
Dichaea picta Rchb.f.
Dichaea sp.
Dimerandra elegans (Focke) Siegerist
Elleanthus caravata (Aubl.) Rchb.f.
Elleanthus graminifolius (Barb. Rodr.) Løjtnant
Encyclia aemula (Lindl.) Carnevali & Ramirez
Encyclia calamaria (Lindl.) Pabst
Encyclia diurna (Jacq.) Schltr.
Encyclia oncidioides (Lindl.) Schltr.
Encyclia pygmaea (Hook.) Dressler
Encyclia selligera (Batem. ex Lindl.) Schltr.
Encyclia vespa (Vell.) Dressler
Epidendrum anceps Jacq.
Epidendrum carpophorum Barb. Rodr.
Epidendrum compressum Griseb.
Epidendrum difforme Jacq.
Epidendrum diffusum Sw. var ?
Epidendrum microphyllum Lindl.
Epidendrum nocturnum Jacq.
Epidendrum purpurascens Focke
Epidendrum rigidum Jacq.
Epidendrum schlechterianum Ames
Epidendrum schomburgkii Lindl.
Epidendrum sculptum Rchb.f. var. sculptum
Epidendrum strobiliferum Rchb.f.
Epidendrum sp.
Eulophia alta (L.) Fawc. & Rendle
Gongora atropurpurea Hook.
Gongora quinquenervis Ruiz & Pav.
Jacquiniella globosa (Jacq.) Schltr.
Koellensteinia carraoënsis Garay & Dunsterv.
Koellensteinia graminea (Lindl.) Rchb.f.
Lepanthes helicocephala Rchb.f.
Lockhartia imbricata (Lam.) Hoehne
Macroclinium mirabile (C. Schweinf.) Dodson
Macroclinium wullschlaegelianum (Focke) Dodson
Masdevallia minuta Lindl.
Maxillaria alba (Hook.) Lindl.
Maxillaria caespitifica Rchb.f.
Maxillaria camaridii Rchb.f
Maxillaria desvauxiana Rchb.f.
Maxillaria discolor (Lodd. ex Lindl.) Rchb.f.
Maxillaria imbricata Barb. Rodr.
Maxillaria parkeri Hook.
Maxillaria ponerantha Rchb.f.
Maxillaria rufescens Lindl.
Maxillaria splendens Poepp. & Endl.
Maxillaria superflua Rchb.f.
Maxillaria trinitatis Ames
Maxillaria uncata Lindl.
Maxillaria violaceopunctata Rchb.f.
Maxillaria sp.
Notylia sagittifera (Kunth) Link., Klotzsch & Otto
Octomeria brevifolia Cogn.
Octomeria deltoglossa Garay
Octomeria grandiflora Lindl.
Octomeria minor C. Schweinf.
Octomeria surinamensis Focke
Octomeria sp.
Oncidium lanceanum Lindl.
Oncidium orthostates Ridl.
Orleanesia amazonica Barb. Rodr.
Ornithocephalus bicornis Lindl.
Palmorchis pubescens Barb. Rodr.
Palmorchis sp.

Platystele ovalifolia (Focke) Garay & Dunsterv.
Plectrophora iridifolia Focke
Pleurothallis aristata Hook.
Pleurothallis barbulata Lindl.
Pleurothallis biglandulosa Schltr.
Pleurothallis brevipes Focke
Pleurothallis breviscapa C. Schweinf.
Pleurothallis ciliolata Schltr.
Pleurothallis determannii Luer
Pleurothallis ephemera Lindl.
Pleurothallis glandulosa Ames
Pleurothallis minima C. Schweinf.
Pleurothallis miqueliana (Focke) Lindl.
Pleurothallis monophylla (Hook.) Fawc. & Rendle
Pleurothallis picta Lindl.
Pleurothallis polygonoides Griseb.
Pleurothallis pruinosa Lindl.
Pleurothallis semperflorens Lindl.
Pleurothallis seriata Lindl.
Pleurothallis spiculifera Lindl.
Pleurothallis sp.
Polystachya flavescens (Blume) J.J. Sm.
Polystachya foliosa (Hook.) Rchb.f.
Polystachya stenophylla Schltr.
Quekettia microscopica Lindl.
Quekettia vermeuleniana Determann
Reichenbachanthus reflexus (Lindl.) Brade
Rodriguezia candida (Lindl.) Christenson
Rodriguezia lanceolata Ruiz & Pav.
Rudolfiella aurantiaca (Lindl.) Hoehne
Scaphyglottis modesta (Rchb.f.) Schltr.
Scaphyglottis prolifera Cogn.
Scaphyglottis sickii Pabst
Scaphyglottis stellata Lodd. ex Lindl.
Scaphyglottis violacea Lindl.
Scaphyglottis sp.
Selenipedium palmifolium (Lindl.) Rchb.f.
Stelis argentata Lindl.
Stelis sp.
Trichosalpinx memor (Rchb.f.) Luer
Trichosalpinx orbicularis (Lindl.) Luer
Trigonidium acuminatum Batem.
Trigonidium obtusum Lindl.
Vanilla cristato-callosa Hoehne cf.
Vanilla sp.
Wullschlaegelia calcarata Benth.
Xerorchis sp.

OXALIDACEAE
Oxalis frutescens L.

PALMAE
Astrocaryum gynacanthum Mart.
Bactris balanophora Spruce
Bactris humilis (Wallace) Burret
Bactris oligoclada Burret
Bactris ptariana Steyerm.
Bactris sp.
Desmoncus macroacanthos Mart.
Euterpe oleracea Mart.
Geonoma maxima (Poit.) Kunth var. ambigua (Spruce) Henderson
Geonoma sp.
Jessenia bataua (Mart.) Burret subsp. oligocarpa (Griseb. & H. Wendl.) Balick
Mauritia flexuosa L.f.
Maximiliana maripa (Corrêa) Drude
Oenocarpus bacaba Mart.

PASSIFLORACEAE
Dilkea wallisii Mast. cf.
Passiflora auriculata Kunth
Passiflora cirrhiflora A. Juss.
Passiflora coccinea Aubl.
Passiflora fuchsiiflora Hemsl.
Passiflora garckei Mast.
Passiflora glandulosa Cav.
Passiflora laurifolia L.
Passiflora misera Kunth
Passiflora vespertilio L.
Passiflora sp.

PHYTOLACCACEAE
Phytolacca rivinoides Kunth & Bouché

PIPERACEAE
Peperomia glabella (Sw.) A. Dietr.
Peperomia macrostachya (Vahl) A. Dietr. var. macrostachya
Peperomia maguirei Yunck.
Peperomia obtusifolia (L.) A. Dietr.
Peperomia ouabianae C. DC. var. ouabianae
Peperomia pellucida (L.) Kunth
Peperomia quadrangularis (Thomps.) A. Dietr.
Peperomia rotundifolia (L.) Kunth var. rotundifolia
Peperomia serpens (Sw.) Loudon
Peperomia sp.
Piper arboreum Aubl.
Piper bartlingianum (Miq.) C. DC.
Piper demeraranum (Miq.) C. DC.
Piper hispidum Sw.
Piper hostmannianum (Miq.) C. DC.
Piper insipiens Trel. & Yunck.
Piper nigrispicum C. DC.
Piper trichoneuron (Miq.) C. DC.
Piper sp.

POLYGALACEAE
Bredemeyera densiflora A.W. Benn. var. glabra A.W. Benn.
Moutabea guianensis Aubl.
Securidaca diversifolia (L.) S.F. Blake cf.
Securidaca paniculata Rich. var. lasiocarpa Oort
Securidaca spinifex Sandw.

POLYGONACEAE
Coccoloba gymnorrachis Sandw.
Coccoloba lucidula Benth. cf.
Coccoloba marginata Benth.
Coccoloba parimensis Benth.
Coccoloba striata Benth.
Symmeria paniculata Benth.

PORTULACACEAE
Talinum fruticosum (L.) Juss.

PROTEACEAE
Panopsis sessilifolia (Rich.) Sandw.

PTERIDOPHYTA
Actinostachys pennula (Sw.) Hook.
Adiantum sp.
Anthrophyum guayanense Hieron.
Anthrophyum sp.
Asplenium angustum Sw.
Asplenium auritum Sw. var. auritum
Asplenium serratum L.
Blechnum serrulatum Rich.
Cochlidium furcatum (Hook. & Grev.) C. Chr.
Cochlidium serrulatum (Sw.) L.E. Bishop
Cochlidium sp.
Cyathea surinamensis (Miq.) Domin
Cyathea sp.
Cyclodium inerme (Fée) A.R. Sm.
Cyclodium meniscioides (Willd.) K. Presl var. meniscioides
Cyclopeltis semicordata (Sw.) Sm.
Dicranoglossum desvauxii (Klotzsch) Proctor
Elaphoglossum glabellum Sm.
Elaphoglossum laminarioides (Bory ex Fée) T. Moore
Elaphoglossum luridum (Fée) Christ.
Elaphoglossum sp.
Grammitis blanchetii (C. Chr.) A.R. Smith
Grammitis flabelliformis (Poir.) C.V. Morton
Hecistopteris pumila (Spreng.) Sm.
Hymenophyllum sp.
Lindsaea divaricata Klotzsch
Lindsaea lancea (L.) Bedd. var. falcata (Dryand.) Rosenst.
Lindsaea sagittata (Aubl.) Dryand.
Lindsaea ulei Hieron.
Lindsaea sp.
Lygodium volubile Sw.
Metaxya rostrata (Kunth) K. Presl
Microgramma lycopodioides (L.) Copel.
Microgramma reptans (Cav.) A.R. Sm.
Microgramma sp.
Nephrolepis biserrata (Sw.) Schott
Nephrolepis multiflora (Roxb.) Jarrett ex C.V. Morton
Pityrogramma calomelanos (L.) Link
Pleopeltis percussa (Cav.) Hook. & Grev.
Polypodium triseriale Sw.
Pteridium aquilinum (L.) Kuhn subsp. caudatum (L.) Bonap.
Schizaea fluminensis Miers ex Sturm
Schizaea incurvata Schkuhr
Schizaea sp.
Tectaria plantaginea (Jacq.) Maxon var. macrocarpa (Fée) C.V. Morton
Thelypteris abrupta (Desv.) Proctor
Thelypteris opulenta (Kaulf.) Fosberg
Trichomanes ankersii Parker ex Hook. & Grev.
Trichomanes arbuscula Desv.
Trichomanes botryoides Kaulf.
Trichomanes diaphanum Kunth
Trichomanes jenmanii Lellinger
Trichomanes kapplerianum J.W. Sturm
Trichomanes martiusii K. Presl
Trichomanes pedicellatum Desv.
Trichomanes pinnatum Hedw.
Trichomanes punctatum Poir. subsp. labiatum (Jenm.) W. Boer
Trichomanes sp.
Triplophyllum acutilobum Holttum
Vittaria costata Kunze
Vittaria lineata (L.) Sm.
Vittaria sp.

QUIINACEAE
Quiina guianensis Aubl.
Quiina indigofera Sandw.
Quiina obovata Tul. cf.
Touroulia guianensis Aubl.

RAPATEACEAE
Rapatea paludosa Aubl. var. paludosa

RHABDODENDRACEAE
Rhabdodendron amazonicum (Spruce ex Benth.) Huber

RHAMNACEAE
Gouania blanchetiana Miq.

RHIZOPHORACEAE
Cassipourea guianensis Aubl.
Cassipourea lasiocalyx Alston

ROSACEAE
Prunus myrtifolia (L.) Urb.

RUBIACEAE
Amaioua guianensis Aubl. var. guianensis
Borreria capitata (Ruiz & Pav.) DC.
Coccocypselum guianense (Aubl.) K. Schum.
Coccocypselum hirsutum Bartl. ex DC.
Cosmibuena grandiflora (Ruiz & Pav.) Rusby var. grandiflora
Diodia ocimifolia (Willd. ex Roem. & Schult.) Bremek.

Duroia amapana Steyerm.
Duroia eriopila L.f. var. eriopila
Duroia hirsuta (Poepp. & Endl.) K. Schum.
Duroia sp.
Faramea occidentalis (L.) A. Rich. subsp. occidentalis
Faramea quadricostata Bremek.
Faramea sessilifolia (Kunth) A. DC.
Ferdinandusa goudotiana K. Schum.
Genipa americana L.
Geophila cordifolia Miq. var. cordifolia
Gonzalagunia dicocca Cham. & Schltdl.
Hillia illustris (Vell.) K. Schum.
Isertia hypoleuca Benth.
Isertia sp.
Ixora aluminicola Steyerm.
Ixora davisii Sandw.
Ixora floribunda (A. Rich.) Griseb.
Ixora graciliflora Benth.
Ixora intropilosa Steyerm.
Malanea hypoleuca Steyerm.
Malanea macrophylla Bartl. ex Griseb. var. macrophylla
Malanea sarmentosa Aubl.
Morinda brachycalyx (Bremek.) Steyerm.
Morinda debilis (Sandw.) Steyerm.
Morinda tenuiflora (Benth.) Steyerm.
Oldenlandia corymbosa L.
Pagamea capitata Benth. subsp. thibaudiifolia (Wernham) Steyerm.
Pagamea guianensis Aubl.
Palicourea crocea (Sw.) Roem. & Schult.
Palicourea croceoides Ham.
Palicourea guianensis Aubl. subsp. occidentalis Steyerm.
Palicourea riparia Benth.
Palicourea sp.
Posoqueria latifolia (Rudge) Roem. & Schult.
Psychotria anceps Kunth
Psychotria apoda Steyerm.
Psychotria astrellantha Wernham
Psychotria bahiensis DC. var. bahiensis
Psychotria capitata Ruiz & Pav. subsp. inundata (Benth.) Steyerm.
Psychotria cuspidata Bredem. ex Roem. & Schult.
Psychotria deflexa DC.
Psychotria glomerulata (Donn. Sm.) Steyerm.
Psychotria hoffmannseggiana (Willd. ex Roem. & Schult.) Muell. Arg. var. hoffmannseggiana
Psychotria irwinii Steyerm. cf.
Psychotria lupulina Benth. subsp. lupulina
Psychotria mapourioides DC. var. chionantha (DC.) Steyerm.
Psychotria officinalis (Aubl.) Sandw. subsp. officinalis
Psychotria puberulenta Steyerm.
Psychotria rosea (Benth.) Muell. Arg.
Psychotria uliginosa Sw.
Psychotria sp.
Randia asperifolia (Sandw.) Sandw.
Rudgea cornifolia (Kunth) Standl. cf.
Rudgea cornigera Bremek.
Rudgea hostmanniana Benth. subsp. hostmanniana
Sabicea aspera Aubl.
Sabicea glabrescens Benth.
Sabicea surinamensis Bremek.
Sipanea biflora (L.f.) Cham. & Schltd.
Sipanea pratensis Aubl.
Spermacoce capitata Ruiz & Pav.
Uncaria guianensis (Aubl.) J.F. Gmel.

RUTACEAE
Angostura toxicaria (Spruce ex Engl.) Albuq.
Hortia regia Sandw.
Ticorea pedicellata DC.
Zanthoxylum apiculata (Sandw.) Waterman

SAPINDACEAE
Cupania hirsuta Radlk.
Cupania scrobiculata Rich.
Cupania sp.
Matayba arborescens (Aubl.) Radlk.
Matayba camptoneura Radlk.
Matayba inelegans Spruce ex Radlk.
Matayba laevigata (Miq.) Radlk.
Matayba macrostylis Radlk.
Matayba opaca Radlk.
Matayba sp.
Paullinia capreolata (Aubl.) Radlk.
Paullinia grandifolia Benth. ex Radlk.
Paullinia ingaefolia Rich.
Paullinia livescens Radlk. cf.
Paullinia pachycarpa Radlk.
Paullinia pinnata L.
Paullinia rufescens Rich. ex Juss.
Paullinia rugosa Benth. ex Radlk.
Paullinia sp.
Serjania paucidentata DC.
Talisia elephantipes Sandw.
Talisia furfuracea Sandw.
Talisia microphylla Uittien
Talisia squarrosa Radlk.
Talisia sp.
Toulicia elliptica Radlk.
Toulicia guianensis Aubl.
Toulicia pulvinata Radlk.

SAPOTACEAE
Chrysophyllum lucentifolium Cronquist var. pachycarpum Pires & T.D. Penn.
Chrysophyllum pomiferum (Eyma) T.D. Penn.
Chrysophyllum sanguinolentum (Pierre) Baehni
Chrysophyllum sparsiflorum Klotzsch ex Miq.
Ecclinusa guianensis Eyma cf.
Ecclinusa psilophylla Sandw.

Manilkara bidentata (A. DC.) Chev. subsp. surinamensis (Miq.) T.D. Penn.
Micropholis venulosa (Mart. & Eichler) Pierre
Micropholis sp.
Pouteria ambelaniifolia (Sandw.) T.D. Penn.
Pouteria caimito (Ruiz & Pav.) Radlk.
Pouteria cladantha Sandw.
Pouteria coriacea (Pierre) Pierre
Pouteria cuspidata (A. DC.) Baehni subsp. dura (Eyma) T.D. Penn.
Pouteria durlandii (Standl.) Baehni
Pouteria egregia Sandw.
Pouteria engleri Eyma
Pouteria filipes Eyma
Pouteria fimbriata Baehni
Pouteria guianensis Aubl.
Pouteria hispida Eyma
Pouteria reticulata (Engl.) Eyma subsp. surinamensis T.D. Penn.
Pouteria speciosa (Ducke) Baehni
Pouteria surumuensis Baehni
Pouteria sp.
Pouteria torta (Mart.) Radlk. subsp. glabra T.D. Penn.
Pouteria venosa (Mart.) Baehni subsp. amazonica T.D. Penn.
Pradosia ptychandra (Eyma) T.D. Penn.
Pradosia schomburgkiana (DC.) Cronquist subsp. schomburgkiana
Pradosia surinamensis (Eyma) T.D. Penn.

SELAGINELLACEAE
Selaginella parkeri (Hook. & Grev.) Spring
Selaginella porelloides (Lam.) Spring
Selaginella producta Baker
Selaginella sp.

SCROPHULARIACEAE
Bacopa aquatica Aubl.
Lindernia crustacea (L.) F. Muell.
Lindernia diffusa (L.) Wettst.
Scoparia dulcis L.

SIMAROUBACEAE
Picramnia guianensis (Aubl.) Jansen-Jacobs
Simaba cedron Planch.
Simaba guianensis Aubl.
Simaba multiflora A. Juss.
Simarouba amara Aubl.

SMILACACEAE
Smilax cumanensis Willd.
Smilax lasseriana Steyerm.
Smilax latipes Gleason
Smilax poeppigii Kunth cf.
Smilax santaremensis A. DC.
Smilax schomburgkiana Kunth
Smilax syphilitica Willd.
Smilax sp1.
Smilax sp2.

SOLANACEAE
Markea coccinea Rich.
Markea sessiliflora Ducke
Markea sp.
Solanum adhaerens Roem. & Schult.
Solanum coriaceum Dunal
Solanum crinitum Lam.
Solanum leucocarpon Rich. ex Dunal
Solanum paludosum Moric.
Solanum rubiginosum Vahl
Solanum rugosum Dunal
Solanum stramoniifolium Jacq.
Solanum subinerme Jacq.
Solanum sp.
Solanum sp. nov.

STERCULIACEAE
Sterculia guianensis Sandw.
Sterculia pruriens (Aubl.) K. Schum. var. ?
Sterculia rugosa R. Brown
Waltheria indica L.

SYMPLOCACEAE
Symplocos sp.

THEACEAE
Ternstroemia punctata (Aubl.) Sw.

THURNIACEAE
Thurnia sphaerocephala (Rudge) Hook.f.

TILIACEAE
Apeiba petoumo Aubl.
Lueheopsis rosea (Ducke) Burret
Lueheopsis rugosa (Pulle) Burret
Mollia lepidota Spruce ex Benth. subsp. sphaerocarpa (Gleason) Meijer
Vasivaea alchorneoides Baill.

TRIGONIACEAE
Trigonia laevis Aubl. var. laevis
Trigonia laevis Aubl. var. microcarpa (Sagot ex Warm.) Sagot
Trigonia sp.

TRIURIDACEAE
Peltophyllum luteum Gardner
Sciaphila albescens Benth.
Triuris hyalina Miers

TURNERACEAE
Turnera aurantiaca Benth.
Turnera rupestris Aubl. var. frutescens (Aubl.) Urb.

ULMACEAE
Trema micrantha (L.) Blume

URTICACEAE
Laportea aestuans (L.) Chew

VERBENACEAE
Aegiphila laevis (Aubl.) J.F. Gmel.
Aegiphila macrantha Ducke
Aegiphila villosa (Aubl.) J.F. Gmel.
Petrea macrostachya Benth.
Petrea volubilis L.

Stachytarpheta jamaicensis (L.) Vahl
Vitex compressa Turcz.
Vitex stahelii Moldenke

VIOLACEAE
Amphirrhox longifolia (St. Hil.) Spreng.
Paypayrola guianensis Aubl.
Paypayrola longifolia Tul.
Paypayrola sp.
Rinorea brevipes (Benth.) S.F. Blake
Rinorea endotricha Sandw.
Rinorea guianensis Aubl.
Rinorea riana Kuntze

VISCACEAE
Phoradendron crassifolium (Pohl ex DC.) Eichler
Phoradendron perrottetii (DC.) Eichler
Phoradendron piperoides (Kunth) Trel.
Phoradendron racemosum (Aubl.) Krug & Urb.
Phoradendron sp.

VITACEAE
Cissus erosa Rich.
Cissus sicyoides L.

VOCHYSIACEAE
Ruizterania albiflora (Warm.) Marcano-Berti
Vochysia schomburgkii Warm.
Vochysia surinamensis Stafleu var. surinamensis

ZINGIBERACEAE
Costus claviger Benoist
Renealmia orinocensis Rusby
Renealmia sp.

CHAPTER 3

Vertical distribution and associations of vascular epiphytes in four different forest types in the Guianas

Renske C. Ek, Hans Ter Steege & Koos C. Biesmeijer

Abstract

The diversity of vascular epiphytes diversity of four neotropical lowland rain forest types in Guyana and French Guiana, South America, was compared in terms of species composition and vertical distribution on host trees. Each forest type showed to have its own site specific epiphytic community. However, in all four forest types we found the same few species dominating the epiphytic vegetation. Furthermore, 67 epiphytic species that occurred in more than three forest types showed comparable vertical distribution patterns in each forest type. Similarities of the four forest types were mainly found within the canopy zones, and these zones also showed the highest species richness. Within each forest type, four groups of major epiphytic habitats, each with its own functional group of epiphytic species, could be distinguished. The four epiphytic habitats with their species group are: A) the tree base, with a group of shade tolerant species, B) the upper trunk, characterized by the absence of epiphytes, C) the inner canopy with the largest group of species having a preference for this habitat, and D) the outer canopy, with a group of xeromorphic species largely restricted to the outer twigs. Furthermore, a group of indifferent species was found, occurring throughout the whole canopy. Differences in species richness between forest types seems to be related to humidity conditions and host specific factors, such as facilitation of the bark and tree architecture.

Introduction

The Guianas (French Guiana, Suriname, and Guyana) are among the few remaining areas of the world largely covered by virgin tropical lowland rain forests. The three countries are part of the Guayana Shield, and belong phytogeographically to the same province (Mori & Prance 1987; Berry *et al.* 1995). Species richness in the Guianas is currently estimated at about 10.000 species of vascular plants (Lindeman & Mori 1989), of which around 9200 species have been described presently (Boggan *et al.* 1997). Richness in epiphytic species contributes to a large extent to the great species diversity in the Neotropics (Gentry & Dodson 1987, 1988). It is generally believed that, although lowland Amazonian forest is comparatively poor in epiphytic species, local epiphytic richness may be as high as, or higher than local epiphytic richness in Andean sites (Gentry 1990). For the Guianas Boggan *et al.* (1997) listed c. 950 vascular species (>10%) occurring in epiphytic genera.

It has been documented that different epiphytic vegetations can be found in different lowland rain forest types (Sandford 1969; Johansson 1974; Ter Steege & Cornelissen 1989), but our knowledge of site-specifity, and geographical distribution of these vegetations is very incomplete.

Within a specific forest type, the species composition and richness of epiphytes differs at different levels in the host tree (*e.g.*, Pittendrigh 1948; Johannson 1974; Longman & Jénik 1987; Ter Steege & Cornelissen 1989; Catling & Lefkovich 1989; Freiberg 1996). This vertical distribution of epiphytes is mainly determined by the variability in microhabitats, characterized by moisture availability, irradiance, available nutrients, and suitable substratum in the canopy and along the trunks (*e.g.*, Johansson 1974; Bennett 1986; Wolf 1994; Benzing 1990, 1995). Lowland tropical forest canopies are typified by intense sunlight, great amplitudes in relative humidity, severe water stress, and a small, pulse-wise supplied pool of nutrients. (*e.g.*, Nadkarni 1988; Benzing 1990; Coxson 1991, 1992; Parker 1995). Colonization of substratum in this patchy environment depends on the seed availability, germination success, and seedling survivalship. Germination success and seedling survivalship depend on the suitability of the substratum and on abiotic factors (moisture, light, and nutrient levels) (Bennett 1986; Benzing 1990). Suitability of the substratum in its turn depends *e.g.* on roughness of the bark, pH, and stability and age of the substratum (*e.g.*, Johansson 1974; Catling *et al.* 1986; Benzing 1990; Wolf 1994).

Host specificity could also play a role in the distribution of epiphytes (Went 1940; Frei 1973; Johansson 1974; Dressler 1981; Ackerman *et al.* 1989; Ter Steege & Cornelissen 1989; Benzing 1990; Wolf 1994). Host-tree specificity is not a direct species-specific phenomenon, but is correlated with the ecological conditions that occur on a specific host-tree species. Host specificity could operate through those factors that directly determine suitability of the substratum (*e.g.* through growth rate of the tree or rates of bark, leaves, and branch shedding), or through species-specific characteristics such as chemical responses, canopy structure and architecture.

This paper aims at comparing the diversity and vertical distribution of vascular epiphytes in four Guianan lowland rain forest types (in Guyana and French Guiana), South America.

The following questions are addressed:

1 What are the patterns of epiphytic species richness in different rain forest types?
2 Are vertical distribution patterns of epiphytic species richness the same in different rain forest types?
3 Can we distinguish groups of epiphytic species that show the same vertical distribution or partition the same environment within a tree (functional groups) in different rain forest types?
4 Can we relate differences in species richness to geographical differences, forest type specific differences, or host-specific differences?

Methods

Study areas

Epiphytes were examined in four different forest types, three near Mabura Hill in the interior of Guyana, 5°20'N, 58°10'W, and one near Saül, in central French Guiana, 5°32'N, 53°12'W.

1 Wallaba forest, Dry evergreen forest, Mabura Hill (WAL DEF).
Wallaba forest occurs on bleached white sand soils (Ter Steege & Cornelissen 1989), and is dominated by *Eperua falcata* Aubl. and *Eperua grandiflora* (Aubl.) Benth. subsp. *guyanensis*. The canopy is regular, without emergent trees, and c. 30 m high. It is a rather light and open forest, with relatively dense undergrowth (Davis & Richards 1932, 1934; Fanshawe 1952). This forest was sampled in 1985.

2 Mora Creek forest, Mabura Hill (MOR CF).
Mora forest is situated on brown sands along creeks and is liable to flooding. *Mora excelsa* Benth. is the dominant tree species and forms the more or less closed main canopy at 40-50 m above ground. Undergrowth is occasional (Fanshawe 1952). This forest was sampled during five months in 1988 and 1989.

3 Greenheart Mixed forest, Mabura Hill (MGH).
During 1994, vascular epiphytes were enumerated in Greenheart forest on brown sands (Ek & Van der Hout, chapter 4). Greenheart forest is a type of mixed tropical rain forest and *Chlorocardium rodiei* (Schomb.) Rohwer, Richter & Van der Werff is one of the dominant tree species, together with Black Kakaralli (*Eschweilera sagotiana* Miers) and Lecythidaceae spp. The forest has a slightly irregular canopy, with occasional to frequent undergrowth (Polak 1992).

4 Mixed forest, Saül (MIX).
In 1986 samples were taken in a mixed forest on lateritic soil, without dominant tree species (de Granville 1988; Mori & Boom 1987; Mori *et al.* 1997). The

canopy was irregular, 30 to 35 m high, with occasional emergents up to 60 m. The inventoried trees were situated near creeks in depressions, between 180 and 210 m altitude. The undergrowth was occasional to frequent (pers. obs.).

Rainfall near Mabura amounts to between 2500 and 3400 mm per year, with an average of 2700 mm (Van Kekem *et al.* 1996). This slightly exceeds the annual rainfall in Saül, which varies from 2000 - 3000 mm per year (de Granville 1988), with an average of 2413 mm (Mori & Prance 1987). Both sites show approximately the same seasonality, with two dry and two wet seasons (Schulz 1960; Mori & Prance 1987; Ter Steege & Persaud 1991).

Data collection and analyses

In *WAL DEF, MOR CF,* and *MIX* standing mature trees were climbed and sampled, using mountaineering techniques (Perry 1981; Ter Steege & Cornelissen 1988). The trees were selected based upon their accessibility and state of health. Trees heavily infested by hemiparasites or largely covered by lianas were avoided. In *WAL DEF* eleven individuals (5 *Eperua grandiflora* and 6 *Eperua falcata),* in *MOR CF* fifteen individuals of *Mora excelsa,* and in *MIX* thirty individuals (with 21 different host-tree species, including 4 *Hura crepitans* L.) were climbed and sampled. In *MGH* vascular epiphytes were enumerated during logging in Greenheart forest (Ek & Van der Hout, chapter 4). In this forest type trees were selected *a priori* based on felling lists. A total number of 105 trees of 11 different host-tree species were sampled just after felling, representing all harvested tree species.

The trees were *a priori* divided into six zones (Johansson 1974; Longman & Jénik 1987), as shown in Fig. 1. Within each height zone at least one sample plot, of at least minimal area, was laid out. Minimal area indicates that the plot was extended till no new species were encountered. As a result for each tree at least 6 plots were sampled. In the canopy zones most plots were placed on upper and lateral sides of a branch, and if needed (*e.g.,* if the species composition between branches differed) a second sample plot was laid out. More details on the methods of sampling can be found in Ter Steege & Cornelissen (1989). In *WAL DEF, MOR CF,* and *MGH* individuals of epiphytes were counted. In *MIX,* however, percentage of cover of vascular epiphytes was recorded. Due to this slight difference in sampling method between sites we used absence/presence data for comparisons between the sites. Hemiepiphytes or facultative epiphytes were included when found rooting within the sampling plots.

Vouchers of specimens of all four studies are deposited in the Utrecht Herbarium (U), in the herbarium of the University of Guyana (BRG) for the Guyanese collections, and in the herbarium of ORSTOM, Cayenne (CAY) and the New York Botanical Garden (NY) for the French Guiana collections. Nomenclature follows Boggan *et al.* (1997).

Only presence or absence of species (binarised data) was used in the analyses. Most analyses were performed with all 142 species of vascular epiphytes that occurred on more than one host-tree. Furthermore, to evaluate the difference in richness between

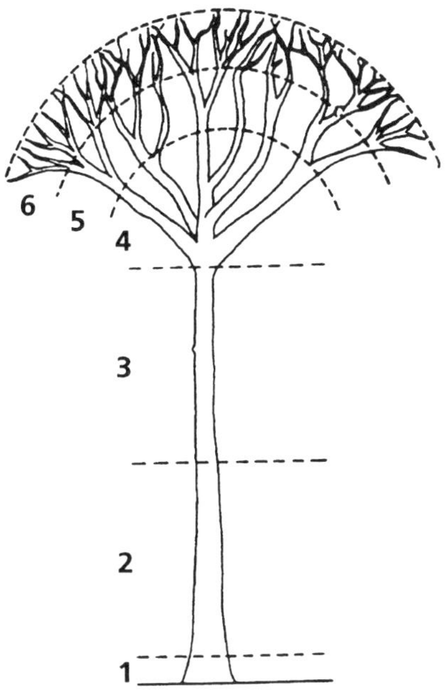

Figure 3.1
Scheme of zones in a rain forest tree. 1= tree base, 2= lower trunk, 3= upper trunk, 4= lower canopy, 5= middle canopy, 6= outer canopy (after Johansson 1974, Longman & Jénik 1987).

the forest types regardless of differences in numbers of sampled trees, randomized cumulative species curves were used. For those forest types in which more than ten trees were sampled, ten trees were randomly selected. The randomized curve was obtained by randomly pooling the trees twenty times.

Floristic data on the level of forest type, host-trees, and tree zones were compared using agglomerative cluster algorithms using BIOPAT (Hogeweg & Hesper 1972, 1981). The similarity matrix used within agglomerative cluster analyses was calculated using mean square distances, and Ward's averaging (minimum increase of the mean sum of squares of distances within clusters (Ward 1963)). Similarity in species diversity between sites was also calculated with the Jaccard's Index (1912), which gives the proportion of species out of the total species list for two samples, which is common to both. Both ways of calculating similarities are used due to the fact that with using mean square distances also the non-presence of species in two samples is considered, while Jaccard considers the proportion of the shared species in relation to the total species richness of both plots. As a result Jaccard's Index emphasizes the species-richness of a sample, while mean square distances emphasizes the species-poorness of a sample. Due to the tendency of Ward's averaging to form clusters with an equal number of members, Group averaging was used as a control. The patterns thus derived were visualized using dendrograms.

To investigate whether or not common species of the four forest types showed the same vertical distribution pattern, we used an agglomerative clustering of tree zone samples of all forest types together. In this analysis we used the most common species,

Table 3.1
Number of epiphytic species in four different forest types in the Guianas.

	N of sampled trees	N of epiphytic species							
		on all trees	on > 1 tree	per tree max	per tree mean	per tree stdev	true	hemi	unique per site
All	161	216	142				190	26	
Wallaba forest (WAL DEF)	11	64	45	37	19.45	11.40	61	3	10
Mora forest (MOR CF)	15	84	76	43	28.50	6.62	77	7	0
Greenheart forest (MGH)	105	93	74	15	5.86	2.98	84	9	22
Mixed forest (MIX)	30	174	97	64	36.70	9.36	152	22	73

meaning those species that occurred in 3 or more sites (N=69). Clusters were based on optimal splitting levels (Hogeweg & Hesper 1981), and characterized using Kruskal Wallis one-way analysis of variance as a measure of importance of the characters. This test indicated which species contributed significantly to the sample plot clusters, and those species were chosen as examples for the species groups.

Results

Floristics

A total number of 216 vascular epiphytic species were found in the four forest types of Guianan lowland rain forest. Full names with authorities of species found, are given in Appendix 1.

Richness in vascular epiphytic species is highest in *MIX* compared with the other three forest types (table 1). Moreover, the maximum number of epiphytic species on one single tree (64) was recorded in *MIX* (64 spp., *MOR CF* : 43, *WAL DEF* : 37, *MGH* : 15). The number of species shared between forest types is lower among the Guyana forest sites than between each of these and the French Guiana site (*MIX*) (table 2). *MIX* and *MOR CF* were most similar as expressed by the Jaccard Index (table 2, appendix 1). All 76 species that occurred in *MOR CF* on more than one tree also occurred in *MIX*.

To evaluate the difference in richness between the forest types regardless of differences in numbers of sampled trees, randomized cumulative species curves were used. The results are presented in Fig. 2. The species richness of the forest types showed the same ranking as according to the maximum number of epiphytic species per tree.

Orchidaceae were the most important group of vascular epiphytes (39% of the total number), followed by Pteridophyta (18%), Araceae (13%), and Bromeliaceae

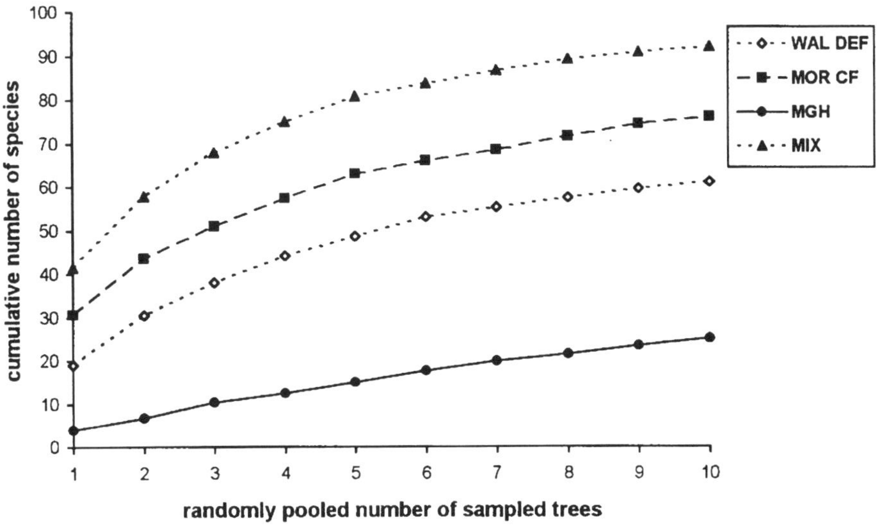

Figure 3.2
Randomized cumulative species curves for four forest types in the Guianas. For each forest type ten trees are used or randomly selected, and the randomized curve is obtained by randomly pooling the trees twenty times.

Table 3.3
Number of epiphytic species per plant family for all species occurring in four forest types in the Guianas.

	WAL DEF		MOR CF		MGH		MIX	
Family	#	%	#	%	#	%	#	%
Orchidaceae	42	64.6	36	47.4	47	52.2	64	36.6
Pteridophyta	5	5.5	13	17.1	14	15.6	35	20.0
Araceae	3	3.3	9	11.8	11	12.2	24	13.7
Bromeliaceae	5	5.5	9	11.8	7	7.8	19	10.9
Cactaceae	1	1.1	1	1.3	3	3.3	3	1.7
Bignoniaceae	1	1.1	3	3.9	1	1.1	5	2.9
Gesneriaceae	1	1.1	1	1.3	2	2.2	4	2.3
Piperaceae	1	1.1	2	2.6	2	2.2	5	2.9
Marcgraviaceae	0	0.0	0	0.0	1	1.1	3	1.7
Guttiferae	1	1.1	0	0.0	1	1.1	5	2.9
Moraceae	0	0.0	0	0.0	1	1.1	0	0.0
Ratio Pter./Orch.	0.12		0.36		0.30		0.55	

Table 3.2
Co-occurrence of epiphytic species and similarity for four different forest types in the Guianas, and the similarity of tree zones between forest types.

Forest types:

Co-occurrence	Total	WAL DEF	MOR CF	MGH
WAL DEF	54			
MOR CF	76	33		
MGH	71	41	41	
MIX	100	45	76	62
Jaccard's Index	**WAL DEF**	**MOR CF**	**MGH**	
MOR CF	0.29			
MGH	0.28	0.32		
MIX	**0.33**	**0.76**	**0.38**	

Tree zones:	WAL DEF						MOR CF		
Jaccard's Index	1	2	3	4	5	6	1	2	3
MOR1	0.00								
MOR2	0.04	0.08							
MOR3	0.02	0.05	0.19						
MOR4	0.02	0.09	0.19	**0.23**					
MOR5	0.02	0.13	0.19	**0.20**	0.16				
MOR6	0.00	0.12	0.05	0.07	0.10	0.14			
GH1	0.00						0.00		
GH2	0.00	0.07					0.00	0.07	
GH3	0.00	0.09	**0.20**				0.09	0.10	0.11
GH4	0.02	0.07	0.19	**0.21**			0.05	0.11	0.14
GH5	0.00	0.02	**0.25**	**0.20**	**0.21**		0.07	0.14	
GH6	0.00	0.00	0.14	0.11	0.09	0.11	0.03	0.07	
MIX1	0.04						0.06		
MIX2	0.05	0.14					0.14	0.11	
MIX3	0.00	0.09	0.19				0.13	**0.26**	**0.42**
MIX4	0.00	0.10	0.16	**0.27**			0.14	**0.29**	**0.48**
MIX5	0.00	0.10	0.17	**0.28**	0.19		0.14	**0.29**	**0.42**
MIX6	0.00	0.09	0.18	**0.27**	**0.22**	**0.26**	0.12	**0.28**	**0.40**

Similarity calculated with the Jaccard Index (1912). Calculations for the forest types with all species, for the tree zones with those species that occurred on more than one tree (N=142). **Bold** values are ≥ 0.20.

(9%) (appendix 1). The relative importance of the taxa differed greatly between the forest types (table 3). Orchids were the most diverse group in all four forest types, and had the highest percentage of total number of species in *WAL DEF.* In the latter forest type, Pteridophyta and Araceae were (relatively) poorly represented. The Pteridophyta to Orchidaceae ratio, considered as a humidity indicator by Johannson (1974), was lowest in *WAL DEF,* and the highest for *MIX* (table 3).

A total of 27 species (12.5%) of epiphytes occurred in all four forest types, and 14 (51.9%) of them are orchids (appendix 1&2).

Vertical distribution

The vertical distribution patterns of epiphytes and their family composition (Fig. 3) showed that the highest species richness occurred in the (inner) canopy (tree zones (3,) 4 and 5) for all four forest types. Orchidaceae and Bromeliaceae had the highest diversity in the canopy (zone (3,)4,5,6); Pteridophyta had the highest diversity in the inner canopy (zone 4), and Araceae were most diverse from the inner canopy to the tree base (zone 1,2,3,4) except in *WAL DEF* (Fig. 3).

Highest similarities, as expressed by the Jaccard Index, between the forest types

Tree zones:	MOR CF			MGH					
Jaccard's Index	**4**	**5**	**6**	**1**	**2**	**3**	**4**	**5**	**6**
MOR1									
MOR2									
MOR3									
MOR4									
MOR5									
MOR6									
GH1									
GH2									
GH3									
GH4	0.15								
GH5	**0.22**	**0.21**							
GH6	0.10	0.09	0.09						
MIX1				0.04					
MIX2				0.05	0.14				
MIX3				0.02	0.07	0.18			
MIX4	**0.54**			0.00	0.04	0.18	**0.23**		
MIX5	**0.49**	**0.46**		0.00	0.02	0.15	**0.25**	**0.27**	
MIX6	**0.48**	**0.44**	**0.22**	0.00	0.03	0.15	**0.27**	**0.31**	0.14

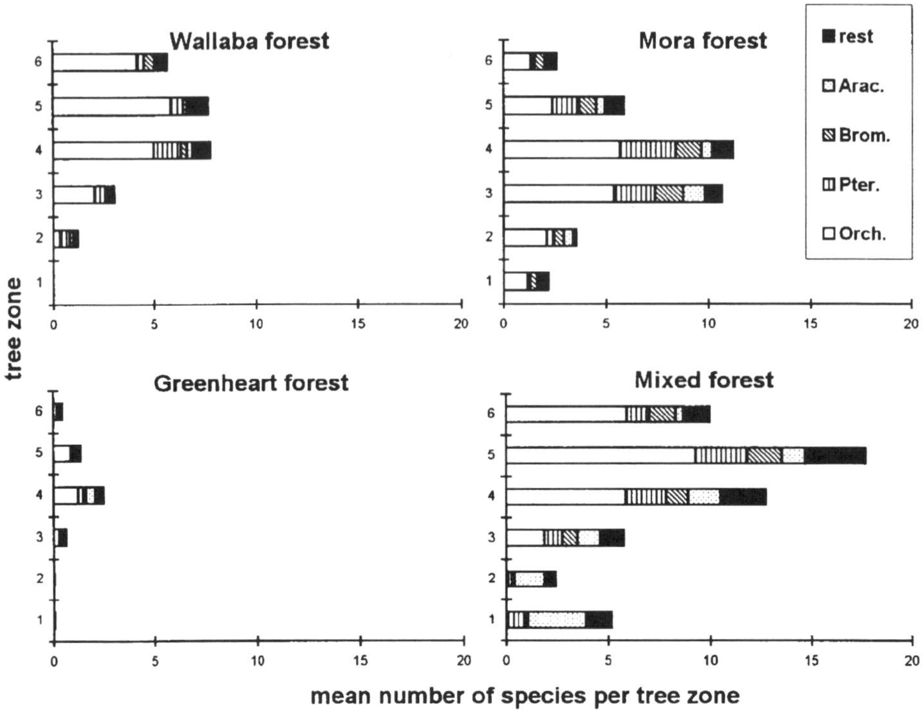

Figure 3.3
Vertical distribution of mean number of epiphytic species per tree for four forest types in the Guianas. For each tree zone the proportional contribution of Orchidaceae, Pteridophyta, Bromeliaceae, Araceae, and others (rest), is indicated.

were found in the canopy zones (zones 4 to 6) (table 2). Especially between the canopy zones of *MOR CF* and *MIX* the highest similarities occur.

Agglomerative clustering of tree zones of all forest types together (N=484) showed that the epiphytic communities were highly site specific: all four forest types clustered out separately, although some similarity existed between *MOR CF* and *MIX* (data not shown). This shows that differences between forest types in site specific species were more important than vertical differentiation within forest types.

However, a number of 69 species occurred in more than 3 forest types. Agglomerative cluster analysis of tree zones using these common species, showed that 4 major epiphytic habitats (A - D) could be distinguished, which are distinguished by their own specific group of species (I - IV) (appendix 2, Fig. 4, Fig. 5):

A A group of sample plots from the tree base (habitat A: zone 1, sometimes extending into zone 2). This cluster comprised sample plots of all four forest types. A group of species (group I) occurred with a preference for this habitat. Examples group I: climbing Araceae, climber spp. (*e.g., Parabignonia steyermarkii*). In *MOR CF* and *MIX* also filmy ferns occurred in this habitat.

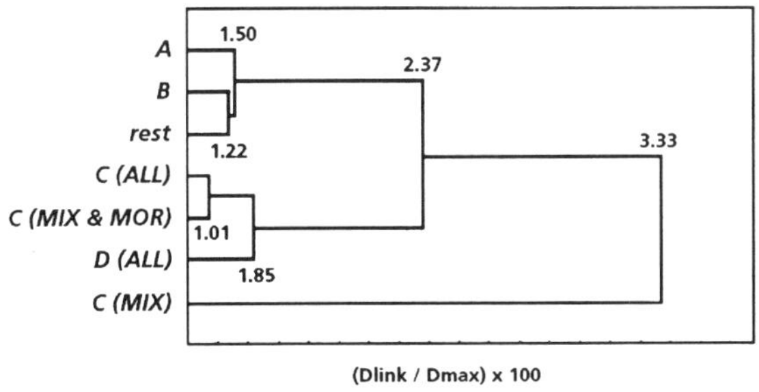

Figure 3.4
Ward's method, Mean square distances:
Diagram of similarities of the different tree habitats (A-D) resulting from agglomerative cluster analyses.

B In this habitat sample plots from the tree trunk (zones 2 and 3) were found, and although in this habitat sometimes climbing species from group I occurred, and sometimes species from group II, this habitat was mainly characterized by the absence of species.

C Habitat C comprised zones (3,) 4, and extends in zone 5. In this habitat the largest group of species was found (group II). This grouping of mainly inner canopy zone sample plots could be found three times within the dendrogram (Fig. 4); one mixed cluster of sampling plots from *MIX, WAL DEF,* and *MGH,* one mixed cluster of sampling plots from *MIX* and *MOR CF*, and one cluster with sampling plots solely from *MIX.* The mixed cluster of *MIX* and *MOR CF* also contained upper trunk (zone 3) sampling plots from *MOR CF*, probably due to irregularities and crotches on the upper trunk of *Mora excelsa.*

Examples group II: *Maxillaria uncata, Anthurium jenmanii,* and *Maxillaria superflua.* Other orchid species such as *Pleurothallis picta, Trichosalpinx orbicularis,* and *Maxillaria violaceopunctata* occurred in this group in all four forest types.

D Habitat D comprised zones 5 and 6. This cluster of outer canopy sampling plots originated from all four forest types. Most species that could be found in this habitat (group III) occurred patchily on the most illuminated twigs.

Examples group III: *Epidendrum strobiliferum* and *Quekettia microscopica.* Other species occurring in this group differed between the four forest types, but were mainly from the genera *Epidendrum s.l.* and *Tillandsia.*

IV Furthermore, a group of species could be found (group IV), that occurred throughout the whole canopy, and occasionally on the tree trunks (habitat A-D).

Examples group IV: *Tillandsia anceps, Codonanthe crassifolia,* and *Stelis argentata.*

The six chosen tree zones sensu Johannson (1974) and Longman & Jénik (1987) were not supported by the occurrence of species groups in neither forest type.

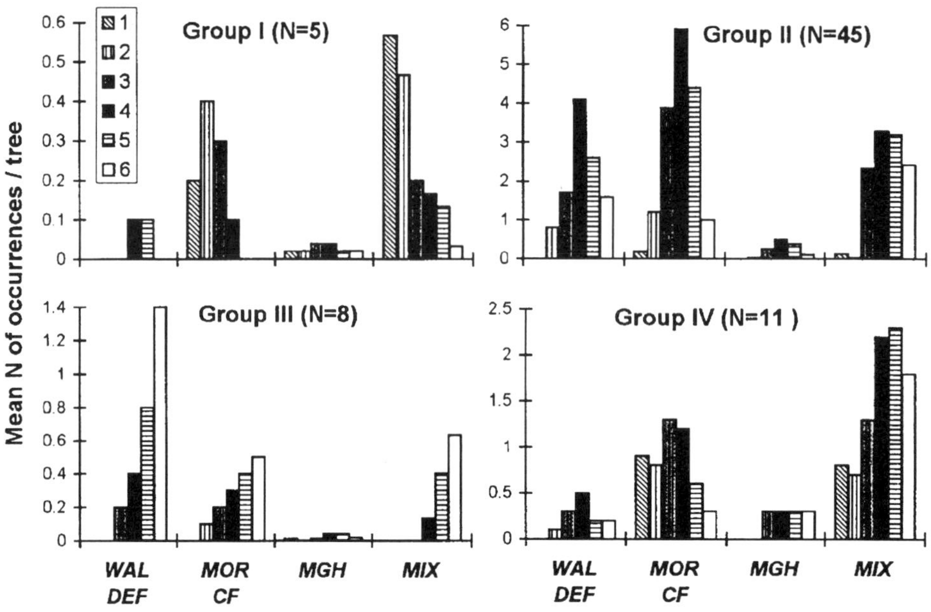

Figure 3.5
Mean number of occurrences per tree zone (1-6) for four species groups in four forest types in the Guianas. For group membership, see appendix 2..

Discussion and conclusions

Species richness

Three investigated types of rain forest, namely Wallaba, Greenheart, and Mixed rain forest in Guyana and French Guiana were found to each contain its own site specific epiphytic community, with distinct differences in species richness (table 1). The epiphytic community of Mora forest, however, seems to belong to the same community as Mixed rain forest. Of the total number of 216 epiphytic species, 142 spp. (66%) appeared on more than one tree. A limited number of 27 species occurred in all four forest types.

The differences in richness between the four forest types (table 1), with Mixed forest as the richest, and Greenheart as the poorest, could be partly due to the larger number of host species sampled in *MIX* (table 1). As the number of epiphytic species increases with number and type of host-trees sampled, it was expected that *MIX* would harbour more epiphytic species than *WAL DEF* and *MOR CF*. *MGH,* however, , with a total of 105 trees enumerated, still remains species poor in relation to *MIX, WAL DEF,* and *MOR CF*. Moreover, the differences in species richness between the forest types remain clear if richness is evaluated on a per tree basis (table 1) or on the basis of randomized cumulative species curves (Fig. 2). So it seems that other factors determine the differences in species richness.

Geographical distribution patterns

If geographical distance would play a role in determining differences in epiphytic species composition between forest types it would be expected that the three forest types near Mabura Hill would have higher similarities in species composition to each other, than to the forest type near Saül. However, shared number of species and Jaccard's Index of similarity between the forest types (table 2), or between specific tree zones (table 2), is always the highest with *MIX* (Saül) for each of the forest types near Mabura Hill. *MOR CF* even shares all its species that occur on more than one tree with *MIX*. This pattern of diversity contrasts with distance dependent patterns of similarity found within liana communities in the Mabura Hill area (Ek & Van der Hout (chapter 4)). This may be a consequence of the long dispersal range of many epiphytic species. For instance orchids, and many ferns, are renowned for their capability of long-distance dispersal, due to dust-sized seeds (Benzing 1990).

Forest type specificity

A second explanation for the differences in epiphytic species richness could be the variation in light and moisture gradients in the forest canopies. The mixed forest in Saül, with a rather open canopy structure, and both emergent and subcanopy trees sampled, has a more variable distribution of light and moisture availability than the forest types with dominant tree species (*WAL DEF, MOR CF, MGH*). Such a heterogeneous canopy leads to a more patchy environment for epiphytes. If strong niche differentiation between epiphytes would exist, this would lead to more different microhabitats for epiphytes. Furthermore, turnover of trees in forests with a heterogeneous canopy is expected to be higher than in more closed canopies (as in *WAL DEF, MOR CF,* and *MGH)* (Phillips *et al.* 1994), and this again increases heterogeneity. In view of the above, *MIX* would have the most heterogeneous canopy, followed by the other type of mixed forest, *MGH,* and the relative open forest type of *WAL DEF.* In this ranking *MOR CF* would form the most homogeneous canopy (homogeneous in the sense that the sampled trees show comparable distribution patterns of light and moisture availability). However, the species richness ranking of *MIX, MOR CF, WAL DEF,* and *MGH* as the poorest forest type, does not support this explanation.

Furthermore, within *MIX* an emergent *Hura crepitans* showed a similar vegetation and vertical distribution pattern as three other, non emergent, *Hura*'s (Ek & Montfoort unpubl. data). These results seem to indicate that differences in variation in light and moisture availability within the canopy also do not necessarily lead to a different epiphytic composition within a forest type.

This could lead to the conclusion that our data do not support the hypothesis that differences in microclimatic (specifically light and moisture availability) environment of the canopy promote differences in species richness between forest types. However, these differences do seem to contribute to the distribution patterns of specific epiphytic species (see further, vertical distribution of epiphytes). Furthermore, the obtained ranking in canopy heterogeneity does coincide with the ranking of forest types in relation to the number of unique species per forest type (table 1). So, although differences in total species numbers were not found to be directly related with heterogeneity of the canopy, it does seem to contribute in part to the species richness.

Another possible factor influencing species richness differences between forest types could be moisture availability at lower levels of the forest structure. The ratio ferns to orchids increases from *WAL DEF* (0.12), via *MGH* (0.30) and *MORA CF* (0.36) to *MIX* (0.55) (table 3). According to Johansson (1974), this ratio correlates with airhumidity, due to the fact that fern diversity correlates positively with airhumidity, and orchids negatively (Benzing 1990). The two wettest forest types (*MOR CF* and *MIX*) show also the highest epiphytic species richness at lower levels in the forest structure (Fig. 3).

Concluding, airhumidity does seem to influence species richness differences between forest types, with the more humid forest types being richer in species. This could lead to a division in more humid, species rich, forest types (*MOR CF* and *MIX*), and dryer, species poor, forest types (*WAL DEF* and *MGH*).

Host specificity

A next possible explanation for the differences of species richness between the forest types could be host-specificity. As already mentioned, a similar vegetation composition and vertical distribution pattern of epiphytes was found on both an emergent, and three non-emergent *Hura crepitans* trees in *MIX* (Ek & Montfoort, unpublished data) supporting the hypothesis that host-specificity could play a role in determining differences in species composition and richness between tree species, and consequently, forest types. Furthermore, Ter Steege & Cornelissen (1989) showed differences in epiphytic communities between two different *Eperua* spp.

Host-specificity could be advanced by species-specific differences in roughness of the bark. However, there was no significant correlation between epiphytic species richness, tree height, or roughness of the bark for the thirty trees measured in *MIX* (Saül) (Ek & Montfoort, unpublished data). These results seem to indicate that within this forest type host specificity as characterized by roughness of the bark does not play a role.

Host-specificity could be expressed by stability and age of the substratum. Although little is known of growth rates of tropical trees, the species poverty of *MGH* could possibly be explained by the fact that within this forest type mainly *Chlorocardium rodiei* carried epiphytes, a species with a flaking bark. Due to this flaking bark the host is inhospitable for long term establishment of epiphytes. This may also explain the low abundance of epiphytes in this forest type: epiphytes always occurred as single individuals, and only when ant gardens were involved, more individuals were encountered (Ek & Van der Hout, chapter 4).

Another host-tree related factor, that could enhance the variability of the canopy environment, is tree architecture (Scatena 1990; Migenis & Ackerman 1993). Tree architecture (branching angles and reiteration patterns) was found to affect patterns of litter fall and stem flow (Scatena 1990), and could therefore influence epiphytic richness and abundance. In the forest types with dominant tree species (*WAL DEF, MOR CF,* and *MGH*) less host-tree species, and therefore less variety of tree architectural types was sampled. If a division in richness classes between humid and dryer forest types is accepted, than the differences found within those classes could possibly be caused by differences in number of sampled architectural types. This host-tree characteristic could also lead to canopy heterogeneity as mentioned under 'forest type specificity', only this time not as a result of heterogeneity of the overall canopy structure, but as heterogeneity between the canopies of individual trees.

Vertical distribution patterns

Among epiphytic species occurring in four forest types in the Guianas comparable vertical distribution patterns of species richness were found (Fig. 3), with the highest species richness in the inner canopy. The inner canopy can be considered a heterogeneous environment, with thick, forking branches, with or without packets of humus in their forks, substrate of different ages, and availability of water and nutrients is likely higher and less pulse supplied than in the outer canopy (Benzing 1995). The outer canopy experiences higher rates of disturbance, and the substratum is in general younger than in the inner canopy (Benzing 1995). These factors probably lead to the difference in species composition and richness between the inner and outer canopy.

In the different forest types in the Guianas three groups of epiphytic species (I - III) were found with distinct preferences for a specific epiphytic habitat (appendix 2). These groups more or less coincide with groups described by Pittendrigh (1948) within bromeliad communities in Trinidad. Pittendrigh distinguished between 'shade tolerant' bromeliad species, occurring on lower parts of the tree, 'sun type' bromeliads that could endure exposure to direct light, but mostly appeared at intermediate levels of insulation and moisture availability, and 'exposure type' bromeliads that occurred on upper, well-illuminated twigs and branches, with prolonged periods of drought. Additionally, we found a fourth, indifferent, group of species. The six a priori chosen tree zones after Johannson (1974) and Longman & Jénik (1987) did not match the observed zonation. This could partly be due to the fact that the boundaries between the four epiphytic habitats (A-D) were variable between the four forest types, depending on airhumidity and insolation. For instance, species preferring the shaded and humid tree base (habitat A) in *MOR CF* and *MIX,* occur only in the shaded and humid inner canopy of *WAL DEF* and *MGH* (appendix 2), indicating that in the latter two forest types the tree base is not so shady and humid. Furthermore, species occurring in the most highly illuminated places of the outer canopy of *MOR CF* and *MIX,* occur in *WAL DEF* and *MGH* in much lower tree zones. This result agrees with the observation that higher insolation levels occur at lower tree height levels in *WAL DEF* and *MGH* (pers. obs.). However, separate analyses of *WAL DEF* (Ter Steege & Cornelissen 1989), *MOR CF* (Biesmeijer & Bleij, unpublished data) and *MIX* (Ek & Montfoort, unpublished data), also showed three or four epiphytic habitats. Our results support the division of the tree in four zones of which three have their own assemblage of epiphytic species as equivalent to those proposed by Pittendrigh (1948), and furthermore described by Ter Steege & Cornelissen (1989). The first zone would correspond with a humid tree base zone (habitat A) in which shade adapted epiphytes appear that mostly grow directly on the bark, and prefer or can endure environments in the tree with the lowest illumination and highest humidity. The second zone would correspond with a upper trunk zone, characterized by the absence of epiphytic species. It seems that depending on the light and moisture availability, and the availability of suitable germination sites (*e.g.* crotches in *MOR CF*) in this second zone, species from group I or group III can be found occurring in this upper trunk zone. This zone can also be bare, and it certainly has no specific species assemblage. The third zone would correspond with the heterogeneous environment of the upper tree trunk and inner canopy zone (habitat C). In this zone both shade adapted species appear in the most shaded inner forks and branches, hemiepiphytes that mainly grow in packages of humus and/or bryophytes, as well as species that can endure sites in direct sun ('sun type', sensu Pittendrigh (1948)). The variability of the environment and wide range of growth forms that appear in habitat C contribute to the amount of species in this group. The last tree zone would correspond with the outer canopy zone (habitat D) in which xeromorphic epiphytes appear that experience high insolation, irregular periods of rain, and large fluctuations in temperature (exposure types, sensu Pittendrigh (1948)). These species root directly on the twigs of the tree.

Differences in species richness between forest types seem to be strongly correlated with humidity conditions, resulting in richer epiphytic species richness in more humid conditions. Within forest types, humidity conditions seem to be correlated with the vertical distribution patterns of epiphytic species on their hosts. Furthermore, the heterogeneity of the canopy, the heterogeneity in architecture of the trees, and the suitability of substratum, all seem to contribute to the species composition and abundance of epiphytes.

Acknowledgements

The authors would like to thank M.J.A. Werger, R. Zagt and P. Hogeweg for comments on an early version of the manuscript, and J.H.C. Cornelissen, D. Montfoort, and E. Bleij for their stimulating discussions and collaboration in the fieldwork. Special thanks are due to all people who assisted in the toiling task of identifying the reference collections: J.C. Lindeman (all families), E. Gouda (Bromeliaceae), A.R.A. Görts-van Rijn (Piperaceae & Guttiferae), A.C. de Roon (Marcgraviaceae) and L.Y.Th. Westra & M. Werkhoven (Orchidaceae). The fieldwork concerning epiphytes was initiated by Dr. S.R. Gradstein and was sponsored by the 'Werkgroep Studiereizen Ontwikkelingslanden (WSO), 'Miquel Fonds', 'Utrechts Universiteits Fonds', the Utrecht University, the Tropenbos Foundation, Wageningen, and the European Union.

Appendix 3.1
Epiphytic species occurring in four forest types in the Guianas

ARACEAE
Anthurium gracile (Rudge) Lindl.: *WAL DEF; MGH; MIX.*
Anthurium jenmanii Engl.: *WAL DEF; MOR CF; MGH; MIX.*
Anthurium pentaphyllum (Aubl.) G. Don: *MIX.*
Anthurium trinerve Miq.: *MOR CF; MGH; MIX.*
Anthurium sp.: *MGH.*
Heteropsis flexuosa (Kunth) Bunting: *MGH.*
Monstera adansonii var. klotzschiana (Schott) Madison: *MIX.*
Monstera adansonii var. laniata (Schott) Madison: *MIX.*
Monstera sect. Marcgraviopsis sp. nov.: *MOR CF; MIX.*
Monstera obliqua Miq.: *MIX.*
Monstera spruceana (Schott) Engl.: *MOR CF; MIX.*
Monstera sp.: *MIX.*
Philodendron acutatum Schott: *MIX.*
Philodendron fragrantissimum (Hook.) Kunth: *MOR CF; MGH; MIX.*
Philodendron guianense Croat & Grayum: *MIX.*
Philodendron insigne Schott: *WAL DEF; MOR CF; MIX.*
Philodendron linnaei Kunth: *MOR CF; MGH; MIX.*
Philodendron pedatum (Hook.) Kunth: *MGH.*
Philodendron ornatum Schott: *MIX.*
Philodendron rudgeanum Schott: *MOR CF; MGH; MIX.*
Philodendron scandens K. Koch & Sello: *MGH; MIX.*
Philodendron surinamense (Schott) Engl.: *MOR CF; MIX.*
Rhodospatha venosa Gleason: *MIX.*
Stenospermation multiovulatum (Engl.) N.E. Brown: *MIX.*
Syngonium podophyllum Schott var. vellozianum (Schott) Croat: *MIX.*
sp1: *WAL DEF; MOR CF; MGH; MIX.*

ARALIACEAE
Oreopanax capitatus (Jacq.) Decne. & Planch. var. capitatus: *MIX.*

BEGONIACEAE
Begonia sp.: *MIX.*

BIGNONIACEAE
Arrabidaea sp.: *MIX.*
Parabignonia steyermarkii Sandw.: *MOR CF; MIX.*
Schlegelia paraensis Ducke: *MOR CF; MIX.*
Schlegelia violacea (Aubl.) Griseb.: *WAL DEF; MOR CF; MGH; MIX.*
Tanaecium nocturnum (Barb. Rodr.) Bureau & K. Schum.: *MIX.*

BROMELIACEAE
Aechmea mertensii (Mey.) Schult.f.: *MOR CF; MIX.*
Aechmea sp.: *MGH.*
Araeococcus micranthus Brongn.: *MIX.*
Catopsis sp.: *MIX.*
Guzmania erythrolepis Brongn. ex Planch.: *MIX.*
Guzmania lingulata (L.) Mez: *WAL DEF; MOR CF; MGH; MIX.*
Guzmania sphaeroidea (André) André ex Mez: *MIX.*
Streptocalyx longifolius (Rudge) Baker: *MGH; MIX.*
Tillandsia adpressiflora Mez: *MIX.*
Tillandsia anceps Lodd.: *WAL DEF; MOR CF; MGH; MIX.*
Tillandsia bulbosa Hook.: *WAL DEF; MOR CF; MGH; MIX.*
Tillandsia fasciculata Sw.: *MOR CF; MGH; MIX.*
Tillandsia kegeliana Mez: *MIX.*
Tillandsia monadelpha (E. Morren) Baker: *MOR CF; MIX.*
Tillandsia spiculosa Griseb. var. stenoglossa (L.B. Sm.) Gouda: *WAL DEF; MOR CF; MGH; MIX.*
Tillandsia tenuifolia L.: *MOR CF; MIX.*
Vriesea amazonica (Baker) Mez: *MOR CF; MIX.*
Vriesea pleiosticha (Griseb.) Gouda: *MIX.*
Vriesea splendens (Brogn.) Lem.: *MIX.*

CACTACEAE
Epiphyllum sp.: *MGH; MIX.*
Rhipsalis baccifera (J. Mill.) W.L. Stearn: *MOR CF; MGH; MIX.*
Selenicereus sp.: *WAL DEF; MGH; MIX.*

CUCURBITACEAE
Cayaponia opthalmica R.E. Schultes: *MIX.*

CYCLANTACEAE
Asplundia rigida (Aubl.) Harl.: *MIX.*
Evodianthus funifer (Poit.) Lindm. subsp. funifer: *MGH.*
Ludovia lancifolia Brongn.: *MIX.*
sp.: *MIX.*

ERIOCAULACEAE
Paepalanthus sp.: *WAL DEF.*

GESNERIACEAE
Codonanthe calcarata (Miq.) Hanst.: *MGH; MIX.*
Codonanthe crassifolia (Focke) Morton: *WAL DEF; MOR CF; MGH; MIX.*
Drymonia psilocalyx Leeuwenberg: *MIX.*
Drymonia sp.: *MIX.*
GUTTIFERAE
Clusia fockeana Miq.: *MIX.*
Clusia panapanari (Aubl.) Choisy: *MIX.*
Clusia platystigma Eyma: *MIX.*
Clusia purpurea (Splitg.) Engl.: *MIX.*
Clusia sp.: *WAL DEF; MGH; MIX.*

LEGUMINOSAE (PAPIL.)
Mucuna sloanei Fawcett & Rendle: *MIX.*

LORANTACEAE
Phoradendron sp.: *WAL DEF.*
Phtirusa sp.: *WAL DEF.*

MARCGRAVIACEAE
Marcgravia cf. coriacea Vahl: *MIX.*
Marcgravia cf. gracilis Sagot: *MIX.*
Norantea guianensis Aubl.: *MGH.*
Souroubea guianensis Aubl. subsp. guianensis: *MIX.*

MORACEAE
Ficus schumacheri (Liebm.) Griseb.: *MIX.*
Ficus sp.: *MGH.*

ORCHIDACEAE
Batemannia colleyi Lindl.: *WAL DEF; MGH.*
Brassavola martiana Lindl.: *WAL DEF; MGH.*
Brassia verrucosa Lindl.: *WAL DEF.*
Campylocentrum micranthum (Lindl.) Rolfe: *MIX.*
Catasetum barbatum (Lindl.) Lindl. var. barbatum: *MGH.*
Cryptharrhena kegelii Rchb.f.: *MIX.*
Dichaea kegelii Rchb.f.: *MOR CF; MIX.*
Dichaea picta Rchb.f.: *WAL DEF; MGH.*
Dichaea sp.: *MGH.*
Dimerandra elegans (Focke) Siegerist: *WAL DEF; MOR CF; MIX.*
Elleanthus caravata (Aubl.) Rchb.f.: *MOR CF; MIX.*
Elleanthus graminifolius (Barb. Rodr.) Loijtnant: *MIX.*
Encyclia aemula (Lindl.) Carnevali & Ramirez: *WAL DEF; MOR CF; MGH; MIX.*
Encyclia diurna Schlechter, s.l.: *WAL DEF; MGH.*
Encyclia pygmaea (Hook.) Dressler: *MOR CF; MIX.*
Encyclia vespa (Vell.) Dressler: *MGH.*
Encyclia sp.: *MIX.*
Epidendrum anceps Jacq.: *MOR CF; MGH; MIX.*
Epidendrum compressum Griseb.: *WAL DEF.*
Epidendrum diffusum Sw.: *WAL DEF.*
Epidendrum microphyllum Lindl.: *WAL DEF; MOR CF; MGH; MIX.*
Epidendrum nocturnum Jacq.: *WAL DEF; MGH.*
Epidendrum oldemanii Christenson: *WAL DEF; MOR CF; MIX.*
Epidendrum ramosum Jacq.: *MIX.*
Epidendrum rigidum Jacq.: *MOR CF; MGH; MIX.*
Epidendrum schlechterianum Ames: *WAL DEF; MGH.*
Epidendrum strobileferum Rchb.f.: *WAL DEF; MGH; MIX.*
Epidendrum sp.: *MIX.*
Jacquiniella globosa (Jacq.) Schltr.: *MGH; MIX.*
Lepanthes sp.: *MIX.*
Lockhartia biserra (L.C. Rich.) Christ. & Garay: *MOR CF.*
Lockhartia imbricata (Lam.) Hoehne: *MOR CF; MIX.*
Masdevallia minuta Lindl.: *MOR CF; MIX.*
Maxillaria alba (Hook.) Lindl.: *MOR CF; MGH; MIX.*
Maxillaria caespitifica Rchb.f. : *WAL DEF.*
Maxillaria camaridii Rchb.f.: *WAL DEF; MOR CF; MIX.*
Maxillaria desvauxiana Rchb.f.: *MOR CF; MIX.*
Maxillaria discolor (Lodd. ex Lindl.) Rchb.f.: *WAL DEF; MOR CF; MIX.*
Maxillaria parviflora (Poepp. & Engl.) Garay: *MOR CF; MIX.*
Maxillaria ponerantha Rchb.f.: *MOR CF; MIX.*
Maxillaria rufescens Lindl.: *WAL DEF; MOR CF; MGH; MIX.*
Maxillaria superflua Rchb.f.: *WAL DEF; MOR CF; MGH; MIX.*

Maxillaria uncata Lindl.: *WAL DEF; MOR CF; MGH; MIX.*
Maxillaria violaceopunctata Rchb.f.: *WAL DEF; MOR CF; MGH; MIX.*
Notylia fragrans Wullschl. ex Focke: *MIX.*
Octomeria brevifolia Cogn.: *MGH.*
Octomeria deltoglossa Garay: *WAL DEF; MGH.*
Octomeria minor C. Schweinf.: *MGH; MIX.*
Octomeria surinamensis Focke: *MGH, MIX.*
Orleanesia amazonica Barb. Rodr.: *WAL DEF.*
Ornithocephalus bicornis Lindl.: *WAL DEF; MIX.*
Platystele ovalifolia (Focke) Garay & Dunsterv.: *WAL DEF; MOR CF; MGH; MIX.*
Plectrophora iridifolia Focke: *WAL DEF; MGH; MIX.*
Pleurothallis archidiaconi Ames: *MOR CF; MIX.*
Pleurothallis aristata Hook.: *MGH; MIX.*
Pleurothallis brevipes Focke: *WAL DEF; MGH.*
Pleurothallis consimilis Ames: *MIX.*
Pleurothallis determannii Luer: *MGH.*
Pleurothallis minima C. Schweinf.: *WAL DEF; MOR CF; MIX.*
Pleurothallis nanifolia Foldats cf.: *MIX.*
Pleurothallis picta Lindl.: *WAL DEF; MOR CF; MGH; MIX.*
Pleurothallis polygonoides Griseb.: *WAL DEF; MOR CF; MGH; MIX.*
Pleurothallis pruinosa Lindl.: *MOR CF; MGH; MIX.*
Pleurothallis semperflorens Lindl.: *MOR CF; MGH; MIX.*
Pleurothallis suspensa Luer: *MIX.*
Pleurothallius glandulosa Ames: *MGH.*
Pleurothallis sp1.: *MIX.*
Pleurothallis sp2.: *MIX.*
Polystachya flavescens (Blume) J.J.Sm.: *WAL DEF.*
Polystachya foliosa (Hook.) Rchb.f.: *MOR CF; MGH; MIX.*
Polystachya stenophylla Schltr.: *MIX.*
Quekettia microscopica Lindl.: *WAL DEF; MOR CF; MGH; MIX.*
Quekettia vermeuleniana Determann: *WAL DEF.*
Reichenbachantus reflexus (Lindl.) Brade: *MGH.*
Rodriguezia lanceolata Ruiz & Pav.: *WAL DEF; MGH; MIX.*
Scaphyglottis fusiformis (Griseb.) Schultes: *MIX.*
Scaphyglottis prolifera Cogn.: *MOR CF; MGH; MIX.*
Scaphyglottis cf. modesta (Rchb.f.) Schltr.: *MGH; MIX.*
Scaphyglottis stellata Lodd. ex Lindl.: *MGH.*
Scaphyglottis violacea Lindl.: *WAL DEF; MGH.*
Scaphyglottis sp1.: *MGH.*
Schomburgkia sp.: *WAL DEF; MIX.*
Sobralia suaveolens Rchb.f.: *MOR CF; MIX.*
Stanhopea grandiflora (Lodd.) Lindl. (cf.): *MIX.*
Stelis argentata Lindl.: *WAL DEF; MOR CF; MGH; MIX.*
Trichosalpinx memor (Rchb.f.) Luer: *WAL DEF; MGH; MIX.*
Trichosalpinx orbicularis (Lindl.) Luer: *WAL DEF; MOR CF; MGH; MIX.*
Trigonidium acuminatum Batem.: *WAL DEF; MGH; MIX.*
Vanilla cristato-callosa Hoehne: *MGH.*
Vanilla sp.: *MGH; MIX.*
sp1&2: *WAL DEF; MOR CF; MIX.*

PIPERACEAE
Peperomia glabella (Sw.) A. Dietr. var. glabella: *MIX.*
Peperomia macrostachya (Vahl.) A. Dietr. var. macrostachya: *MGH; MIX.*
Peperomia magnoliifolia (Jacq.) A. Dietr.: *MIX.*
Peperomia rotundifolia (L.) Kunth var. rotundifolia: *WAL DEF; MOR CF; MGH; MIX.*
Peperomia serpens (Sw.) Loudon: *MOR CF; MIX.*

PTERIDOPHYTA
Anetium citrifolium (L.) Splitg.: *MOR CF; MIX.*
Antrophyum cajennense (Desv.) Spreng.: *MGH; MIX.*
Antrophyum guayanense Hieron.: *MIX.*
Antrophyum lanceolatum (L.) Kaulf.: *MIX.*
Asplenium angustum Sw.: *MGH; MIX.*
Asplenium auritum Sw. var. auritum: *MOR CF; MIX.*
Asplenium juglandifolium Lam.: *MIX.*
Asplenium serratum L.: *MIX.*
Campyloneuron phyllitidis (L.) K. Presl: *MIX.*
Campyloneuron repens (Aubl.) K. Presl: *MIX.*
Cochlidium serrulatum (Sw.) L.E. Bishop: *MIX.*
Dicranoglossum desvauxii (Klotzsch) Proctor: *MGH.*
Elaphoglossum glabellum Sm.: *WAL DEF; MOR CF; MGH; MIX.*
Elaphoglossum herminieri (Bory ex Fée) T. Moore: *MIX.*
Elaphoglossum laminarioides (Bory ex Fée) T. Moore: *MOR CF; MIX.*
Elaphoglossum luridum (Fée) Christ: *WAL DEF; MOR CF; MIX.*

Grammitis blanchetii (C.Chr.) A.R.Sm.:*WAL;MGH; MIX.*
Grammitis rostrata (Hook.) R. & A. Tryon: *MIX.*
Grammitis sp.: *MGH.*
Hecistopteris pumila (Spreng.) J. Sm.: *MIX.*
Huperzia dichotoma (Jacq.) Trevis.: *MOR CF; MGH; MIX.*
Huperzia taxifolia (Sw.) Trevis.: *MIX.*
Hymenophyllum hirsutum (L.) Sw.: *MIX.*
Hymenophyllum polyanthos (Sw.) Sw.: *MIX.*
Lomariopsis japurensis (Mart.) Sm.: *MIX.*
Lygodium volubile Sw.: *MGH.*
Microgramma lycopodioides (L.) Copel.: *WAL DEF; MOR CF; MGH; MIX.*
Microgramma piloselloides (L.) Copel.: *MIX.*
Microgramma reptans (Cav.) A.R. Sm.: *MOR CF; MGH; MIX.*
Microgramma tecta (Kaulf.) Alston: *MIX.*
Pleopeltis percussa (Cav.) Hook. & Grev.: *MOR CF; MGH; MIX.*
Polypodium triseriale Sw.: *WAL DEF; MOR CF; MIX.*
Trichomanes crispum L.: *MIX.*
Trichomanes kapplerianum Sturm: *MIX.*
Trichomanes krausii Hook. & Grev.: *MIX.*
Trichomanes punctatum Poir. subsp. labiatum (Jenm.) W. Boer: *MOR CF; MIX.*
Vittaria costata Kunze: *MOR CF; MGH; MIX.*
Vittaria lineata (L.) J.Sm.: *MOR CF; MGH; MIX.*

RUBIACEAE
Hillia illustris (Vell.) K. Schum.: *WAL DEF; MOR CF; MIX.*

SMILACACEAE
Smilax sp.: *MIX.*

Appendix 3.2

Distribution of epiphytic species over tree height zones in four forest types. Epiphytes are listed that occur in three or more forest types.

Group	forest type	Walaba forest						Mora forest					
	tree zone	1	2	3	4	5	6	1	2	3	4	5	6
	nr plots with epiphytes	15	16	16	18	14	13	9	10	10	10	10	9
I	*Antrophyum cajennense*												
I	*Parabignonia steyermarkii*							1					
I	*Philodendron fragrantissimum*								1	1	1		
I	*Philodendron insigne*				1	1		1	2	1			
I	*Philodendron rudgeanum*								1	1			
II	***Anthurium gracile***				1	1						1	
II	***Anthurium jenmanii***				1				1	2	2	1	
II	*Anthurium trinerve*									1	1	1	
II	*Dimerandra elegans*								1	1			
II	***Elaphoglossum glabellum***		2	2	5	2				3	3	2	
II	*Elaphoglossum laminarioides*									1	1	1	
II	***Encyclia aemula***				1	1	1			1	3	3	
II	*Epidendrum anceps*								1			1	
II	***Epidendrum microphyllum***					1				1	1	1	
II	***Grammitis blanchetii***		1		2					1			
II	***Guzmania lingulata***		1	1	2		2		2	4	2	2	1
II	*Hillia illustris*				1					1	1	2	
II	*Huperzia dichotoma*									1	3	1	1
II	*Maxillaria alba*								1	1	1		
II	*Maxillaria camaridii*			1	2	2	1			1	2	1	1
II	*Maxillaria discolor*				1					1	4	1	
II	***Maxillaria rufescens***				1					1			
II	***Maxillaria superflua***			3	4	3	1			1	2	2	1
II	***Maxillaria uncata***			2	4	3	2				3	2	1
II	*Maxillaria violaceopunctata*				1				1	1	1		
II	***Microgramma lycopodioides***		1	1	2	2	2			1	2	3	1
II	*Microgramma reptans*									1	3	2	
II	***Peperomia rotundifolia var. rotundifolia***			1		1	1					1	1
II	*Philodendron linnaei*							1		2	1		
II	***Platystele ovalifolia***		1		1	2	1					1	
II	*Pleopeltis percussa*											1	
II	*Pleurothallis minima*			1		1	1			1	2	1	
II	***Pleurothallis picta***			1	1						2	1	
II	***Pleurothallis polygonoides***				1	1					1	1	1

Appendix 3.2 (continued)

Species indicated **bold** occur in all four forest types. I-IV: group membership according to agglomerative clustering.

Group	forest type	Walaba forest						Mora forest					
	tree zone	1	2	3	4	5	6	1	2	3	4	5	6
	nr plots with epiphytes	15	16	16	18	14	13	9	10	10	10	10	9
I	*Antrophyum cajennense*		1	1			2						
I	*Parabignonia steyermarkii*							2	2	2	1	2	1
I	*Philodendron fragrantissimum*			1	1	1	1	4	4	2	2		
I	*Philodendron insigne*							4	4	1	2	2	
I	*Philodendron rudgeanum*	1	1					5	4	1			
II	***Anthurium gracile***				1	1	1				4	4	1
II	***Anthurium jenmanii***			1	1			2		1	4	2	
II	*Anthurium trinerve*				1	1	1			1	3	2	1
II	*Dimerandra elegans*										2	1	4
II	***Elaphoglossum glabellum***			1	1					2	4	4	
II	*Elaphoglossum laminarioides*									2	1	1	
II	***Encyclia aemula***				1					2	1	4	1
II	*Epidendrum anceps*				1					2	1	4	4
II	***Epidendrum microphyllum***					1				2	2	3	3
II	***Grammitis blanchetii***			1	1					2	2		
II	***Guzmania lingulata***		1					1	1	2	3	3	1
II	*Hillia illustris*									4	3	3	4
II	*Huperzia dichotoma*					1					4	2	
II	*Maxillaria alba*					1				4	3	3	1
II	*Maxillaria camaridii*								2	4	1	1	
II	*Maxillaria discolor*								2	1	1	2	
II	***Maxillaria rufescens***				1					2	2	1	2
II	***Maxillaria superflua***				1	1				1	3	3	2
II	***Maxillaria uncata***			1	1	1				2	4	4	2
II	*Maxillaria violaceopunctata*				1					1	3	2	1
II	***Microgramma lycopodioides***				1						1	1	1
II	*Microgramma reptans*			1	1	1					1	2	1
II	***Peperomia rotundifolia var. rotundifolia***			1						1	1	3	1
II	*Philodendron linnaei*		1	1				1		1	4	4	2
II	***Platystele ovalifolia***			1							1	2	2
II	*Pleopeltis percussa*				1					4	4	3	2
II	*Pleurothallis minima*				1	1				4	2	1	1
II	***Pleurothallis picta***			1	1	1				2	4	2	2
II	***Pleurothallis polygonoides***					1				2	4	4	4

Group	forest type	Walaba forest						Mora forest					
	tree zone	1	2	3	4	5	6	1	2	3	4	5	6
	nr plots with epiphytes	15	16	16	18	14	13	9	10	10	10	10	9
II	*Pleurothallis pruinosa*								1	3	2	1	
II	*Pleurothallis semperflorens*							1	1	1	1	1	
II	*Polypodium triseriale*								1	2	1		
II	*Polystachya foliosa*												
II	*Scaphyglottis prolifera*											1	
II	***Schlegelia violacea***		1								1	2	1
II	***Tillandsia bulbosa***			1	1	1	2			1	2	1	1
II	*Tillandsia fasciculata*										1		
II	***Tillandsia spiculosa var. stenoglossa***				1		2			1	1	1	
II	*Tillandsia tenuifolia*									1	2	1	
II	*Trichosalpinx memor*			1	2	2							
II	***Trichosalpinx orbicularis***			2	2	1	1			1	3	1	
II	*Vittaria costata*								1				
II	*Vittaria lineata*										1	1	
II	*Vriesea amazonica*								1				
II	***Clusia panapanari***			1	3	2				1	2	1	1
III	*Encyclia diurna*				1	1	4						
III	*Encyclia pygmaea*								1	1			
III	***Epidendrum strobiliferum***			1	1	2	3			1	1	2	2
III	*Jacquiniella globosa*												
III	***Plectrophora iridifolia***					1	2						1
III	***Quekettia microscopica***						1						1
III	*Rodriguezia lanceolata*					2	3						
III	***Trigonidium acuminatum***			1	2	2	1				2	2	1
IV	*Anetium citrifolium*									2	3	1	
IV	*Asplenium auritum* var. *auritum*								1	1			
IV	***Codonanthe crassifolia***			1	2	2	2		1	1		1	
IV	*Elaphoglossum luridum*			1					1	3	3	2	
IV	*Epidendrum difforme*							4					1
IV	*Epidendrum rigidum*							1	5	5	3	1	1
IV	*Peperomia serpens*									1			
IV	*Rhipsalis* sp.							3					
IV	*Selenicereus* sp.				1								
IV	***Stelis argentata***			1	2						3	1	
IV	***Tillandsia anceps***		1					1					1

Group	forest type	Walaba forest						Mora forest					
	tree zone	**1**	**2**	**3**	**4**	**5**	**6**	**1**	**2**	**3**	**4**	**5**	**6**
	nr plots with epiphytes	**15**	**16**	**16**	**18**	**14**	**13**	**9**	**10**	**10**	**10**	**10**	**9**
II	*Pleurothallis pruinosa*					1				4	3	2	2
II	*Pleurothallis semperflorens*				1	1					1	1	2
II	*Polypodium triseriale*									4		2	4
II	*Polystachya foliosa*				1	1	1			2		1	1
II	*Scaphyglottis prolifera*			1	1	1					1	2	3
II	***Schlegelia violacea***				1					2	1	1	4
II	***Tillandsia bulbosa***					1				1	2	2	1
II	*Tillandsia fasciculata*					1					1	1	1
II	***Tillandsia spiculosa var. stenoglossa***						1			2	1	2	1
II	*Tillandsia tenuifolia*									1	2	2	2
II	*Trichosalpinx memor*						1				2		
II	***Trichosalpinx orbicularis***			1	1	1	1			1	2	1	2
II	*Vittaria costata*				1					1	1	4	2
II	*Vittaria lineata*			1	1					4	2	2	1
II	*Vriesea amazonica*										2	2	1
II	***Clusia panapanari***			1	2	1					2	1	
III	*Encyclia diurna*	1											
III	*Encyclia pygmaea*										2	1	4
III	***Epidendrum strobiliferum***						1				2	1	3
III	*Jacquiniella globosa*					1						4	4
III	***Plectrophora iridifolia***				1							4	4
III	***Quekettia microscopica***			1	1	1						1	2
III	*Rodriguezia lanceolata*				1	1	1						1
III	***Trigonidium acuminatum***				1	1						1	1
IV	*Anetium citrifolium*							2				2	
IV	*Asplenium auritum* var. *auritum*							2			4	4	2
IV	***Codonanthe crassifolia***			1	1	1	1			2	2	3	2
IV	*Elaphoglossum luridum*							2		2	1	1	2
IV	*Epidendrum difforme*									2	1	1	
IV	*Epidendrum rigidum*						1				1	2	1
IV	*Peperomia serpens*							2			1	2	
IV	*Rhipsalis* sp.			1					2	1	3	1	4
IV	*Selenicereus* sp.				1	1			2	4	1	2	2
IV	***Stelis argentata***			1	1	1	1		1	2	5	4	2
IV	***Tillandsia anceps***								2	2	2	1	2

CHAPTER 4

Botanical diversity of Greenheart dominated mixed rain forest near Mabura Hill, Guyana

Renske C. Ek & Peter Van der Hout

Abstract

Greenheart (*Chlorocardium rodiei*) is the main commercial timber export product of Guyana (South America). The forest type in which it grows, mixed tropical rain forest on terra firme, is called Greenheart forest after this (co-)dominant tree species. This paper gives a description of the species composition and abundance based on 15 hectare of Greenheart forest. In the unexploited Greenheart forest in Pibiri we found 405 botanical species, ranging between 125 and 170 species per ha. Compared to other studies on species richness of mixed tropical rain forest, the Greenheart forest in Guyana can be considered a mediate species-rich to species-poor tropical rain forest type. Furthermore, the relation between species diversity and logging damage as a result of conventional logging methods was evaluated. All growth forms showed their own responses to logging, at different spatial scales. Lianas were the most responsive to logging-induced changes in the environment and showed the highest relative increase in species number. Undergrowth species (herbs, shrubs, and small palms) were very uniform in species composition over exploited and unexploited forest, with a strong dominance of a few species. The composition of tree species after logging is mainly determined by newly established species after logging. The more overlapping gap and skid trail area created during logging, the more new, invading, species arrive after logging (both trees and lianas). In relation to forest management this would indicate that to maintain the characteristic species composition of Greenheart forest, skidding inside gaps should be kept to a minimum.

Introduction

Logging in Guyana occurs mainly in Greenheart (*Chlorocardium rodiei*) dominated mixed forest. Little is known on the plant species composition and abundance of Greenheart dominated forest. Since Davis and Richards published on their research at Moraballi Creek (1932, 1934) and Fanshawe (1952) made a general description of the forest types for Guyana, hardly any quantitative studies of this forest type were undertaken. Recently, Ter Steege *et al.* (1993), Comiskey *et al.* (1994), and Johnston & Gilman (1995) published on the composition of large woody species of forest plots in Guyana, of which a few contained Greenheart forest.

The number of plant species per sample is considered to be a good first order estimate of local plant diversity ('alpha diversity', sensu Whittaker 1977)(*e.g.* Gentry 1990; Valencia *et al.* 1994; Clinebell *et al.* 1995; Condit *et al.* 1996). The observed number of species in any sample of a species-rich community, however, inevitably underestimates the true number of species present (Chazdon *et al.* in press). Where several samples are repeated at different sites, they can be used to estimate the spatial turnover in species between those sites ('beta diversity', sensu Whittaker 1977) (*e.g.*, Clinebell *et al.* 1995).

In this paper we focus on the alpha and beta diversity of mixed tropical rain forest on brown sand dominated by Greenheart in unexploited and exploited circumstances. The question was raised how the composition and development of the botanical community in Greenheart forest changes as a result of logging and logging-related damage. The effects of traditional logging (with high logging intensities and damage) on the plant species composition and abundances are evaluated, with special reference to the influence of skid trails and gap sizes. We describe the botanical diversity of 15 ha of unexploited Greenheart dominated mixed forest and compare the results with the biodiversity found at other sites which had been logged previously. Differences in species diversity due to traditional logging are evaluated in terms of species richness, abundance and growth form spectra. Additionally, we studied the between-plot and within-plot variation in species composition within unexploited, and exploited, Greenheart forest.

Materials and methods

Research sites

The study has been conducted on the timber concession of Demerara Timbers Ltd. near Mabura Hill (5°13'N, 58°48'W). The concession lies between the Essequibo and Demerara Rivers in central Guyana, approximately 240 km from the coast. Greenheart (*Chlorocardium rodiei*) is Guyana's major commercial timber species and the study location is situated in 'Greenheart dominated mixed forest on brown sand'. This forest type will hereafter be referred to as 'Greenheart forest'. According to Van Kekem *et al.* (1996), all study sites are situated in the same mapping unit: with deep

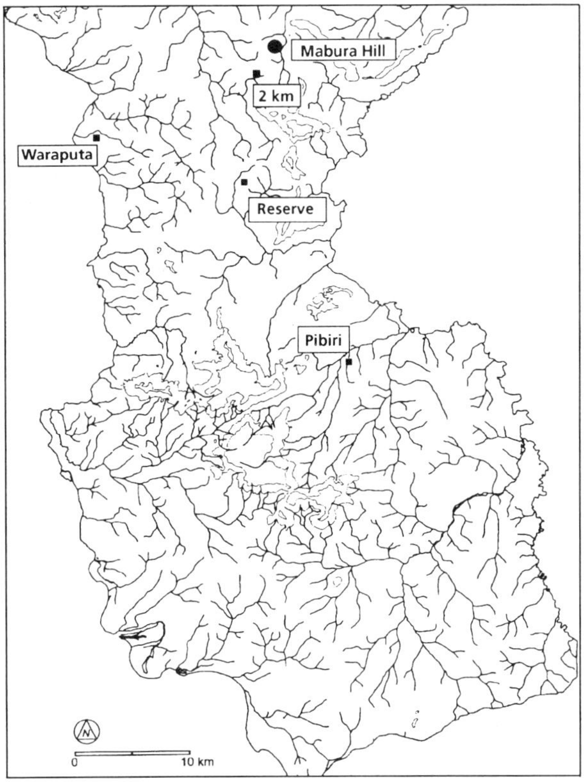

Figure 4.1
Location of the study sites in the vicinity of Mabura Hill, Guyana.

well-drained yellowish brown to strong brown sands in the topsoil and with sandy loams to clay textures in the subsoils. The soils are poor in nutrients and have moderate amounts of available soil moisture. In total 24 hectares of Greenheart forest were inventoried, of which 15 in the West-Pibiri compartment (this site hereafter will be referred to as 'Pibiri'). The enumerations took place during 1993 and 1994. In 1995 three one-ha plots, two logged and one referential, were established in three different exploited sites of Greenheart forest: at 2 km south of Mabura township ('2 km') logged in 1985, in kilometer square no. 34 in the Waraputa compartment ('Waraputa') logged in 1989-1990, and just outside the Mabura Hill Forest Reserve ('Reserve') logged in 1988-1989. The latter three sites were included to investigate the effect of time after logging, and introduced a new level of geographical differences. Information on logging intensities can be found in chapter 5 (table 1).

Nomenclature and growth forms

Nomenclature of botanical species and/or families follows Cronquist (1981), with a few alterations after Maas & Westra (1993). Due to the fact that familial limits are not agreed upon among ferns and fern allies, we follow the common practice to treat the group as one family (Gentry 1990).

Table 4.1

Most abundant species in Greenheart dominated forest, West Pibiri Compartment, Mabura Hill, Guyana. Most abundant species (> 2% of all individuals), freq: frequency (%) of occurrence in the 15 one-ha plots. Canopy trees include all trees with dbh _ 10 cm. Treelets include all trees with a dbh between 5 and 20 cm. Large woody lianas have a dbh > 10 cm. Lianas are all climbers occurring with a height of more than 2 m. Undergrowth includes all herbs, shrubs, and small palms (adult height < 4 m, adult DBH < 5 cm), in this category there is no distinction between undergrowth species occurring with a height of more than 2 meters, and those occurring with a height of more than 50 cm. Epiphytes include all true epiphytes occurring on 105 sampled trees, frequency is not given due to the fact that not all plots are equally sampled.

Large trees		118 sp.	Small trees		140 sp.
Name	% ind	Freq	Name	% ind	Freq
Chlorocardium rodiei	19.9	100.0	*Tapura guianensis*	12.7	100.0
Lecythis confertiflora	12.8	100.0	*Oxandra asbeckii*	9.6	100.0
Catostemma fragrans	8.5	100.0	*Catostemma fragrans*	6.2	100.0
Mora gonggrijpii	3.9	33.3	*Lecythis confertiflora*	4.0	100.0
Carapa guianensis	3.7	66.7	*Licania heteromorpha* var. *perplexans*	3.7	100.0
Eperua falcata	3.3	73.3	*Chlorocardium rodiei*	3.5	86.7
Licania densiflora	3.2	86.7	*Eschweilera sagotiana*	3.4	93.3
Licania alba	3.0	80.0	*Sloanea guianensis*	3.4	60.0
Swartzia leiocalycina	3.0	86.7	*Mora gonggrijpii*	3.3	33.3
Eschweilera sagotiana	2.9	33.3	*Licania canescens*cf.	2.8	86.7
Eschweilera coriacea	2.0	53.3	*Eschweilera wachenheimii*	2.5	46.7
			Paypayrola guianensis	*2.5*	*93.3*
Largewoody lianas		**32 sp.**	**Lianas**		**132 sp.**
Name	% ind	Freq	Name	% ind	Freq
Anomospermum grandifolium	24.4	86.7	*Connarus perrottetii* var. *rufus*	8.2	100.0
Moutabea guianensis	8.9	73.3	*Rourea pubescens* var. *spadicea*	6.5	100.0
Pinzona coriacea	8.1	46.7	*Heteropsis flexuosa*	5.6	100.0
Bauhinia guianensis var. *guianensis*	6.9	46.7	*Maripa scandens*	5.3	100.0
Abuta imene	6.5	33.3	*Odontodenia geminata* cf.	5.2	73.3
Curarea candicans	6.5	13.3	*Connarus megacarpus*	4.7	100.0
Tetracera volubilis subsp. *volubilis*	4.1	53.3	*Lonchocarpus negrensis*	3.6	100.0
Machaerium myrianthum	3.7	46.7	*Coccoloba parimensis*	3.2	100.0
Anemopaegma oligoneuron	2.8	20.0	*Tetracera volubilis* subsp. *volubilis*	2.9	100.0
Doliocarpus guianensis	2.8	26.7	*Pleonotoma albiflora*	2.8	100.0
Machaerium madeirense	2.8	33.3	*Machaerium madeirense*	2.6	100.0
Strychnos melinoniana	2.4	33.3	*Moutabea guianensis*	2.6	100.0
Memora moringifolia	2.0	20.0	*Anomospermum grandifolium*	2.4	100.0
Connarus perrottetii var. *rufus*	2.0	20.0	*Memora racemosa*	2.3	86.7
Cheiloclinium cognatum	2.0	13.3	*Forsteronia guyanensis*	2.2	93.3
			Bauhinia guianensis var. *guianensis*	2.1	60.0
			Pinzona coriacea	2.1	80.0
			Gnetum nodiflorum	2.0	100.0

Epiphytes		65 sp.	Undergrowth		36 sp.
Name	% ind	Freq	Name	% ind	Freq
Codonanthe crassifolia	9.6		*Bactris oligoclada*	43.5	93.3
Maxillaria uncata	6.3		*Bactris balanophora*	10.9	100.0
Anthurium gracile	4.2		*Ischnosiphon puberulus* var. *scaber*	8.3	100.0
Scaphyglottis prolifera	4.2		*Psychotria puberulenta*	7.1	93.3
Codonanthe calcarata	3.3		*Bactris humilis*	5.4	100.0
Anthurium trinerve	3.0		*Geonoma maxima*	3.1	93.3
Gravisia sp.	3.0		*Faramea quadricostata*	2.7	100.0
Brassavola martiana	3.0		*Psychotria apoda*	2.6	93.3
Philodendron fragrantissimum	2.7		*Calyptrocarya glomerulata* var. *glomerulata*	2.4	60.0
Epidendrum sp.	2.4		*Psychotria capitata*	2.3	66.7
Scaphyglottis violacea	2.4		*Voyria corymbosa* subsp. *alba*	2.2	86.7
Stelis argentata	2.4		*Ischnosiphon gracilis*	2.1	86.7
Trichosalpinx orbicularis	2.4				
Philodendron pedatum	2.1				
Batemania colleyi	2.1				
Pleurothallis semperflorens	2.1				
Peperomia macrostachya	2.1				

The growth forms used here are mainly defined after Hubbell & Foster (1990, 1992), although some modifications are made following Gentry and Dodson (1987) in order to group those species, especially the climbers, who could play specific roles in respect to logging operations.

Trees (canopy trees, treelets, and large palms): The tree component includes canopy trees with a diameter at maturity ≥ 20 cm, and trees with a diameter ≥ 5 cm and smaller than 20 cm dbh, and a adult height of more than 4 m. Large palms, occurring in the mentioned size classes, are also included in this category.

The climbers: The climber component includes lianas, woody hemiepiphytes and vines. Lianas are woody climbing plants that rely on other plants for support (Putz 1984b). Woody hemi-epiphytes rely on other woody plants for their support only for a part of their life cycle. Vines are non-woody small tendril climbing herbaceous plants. The term lianas will be used hereafter to refer to all climbers for a matter of convenience. In most research only lianas with a dbh > 10 cm are considered. This size category will be separately evaluated. More detailed effects of logging on the liana flora are treated in Ek *et al.* (submitted, chapter 5).

The undergrowth: herbs, shrubs and small palms: Although neglected in most studies the undergrowth generally contains between 25 to 46 % of the species in wet tropical forests (Gentry & Dodson 1987). In this study we include in the undergrowth category: (A) herbs, terrestrial herbaceous plants and terrestrial woody plants with an adult diameter < 1 cm; (B) shrubs, woody, freestanding plants with an adult height < 4 meters and with a diameter between 1 and 10 cm, and (C) small palms, with an adult height of less than 4 meters and a diameter less than 5 cm.

Epiphytes: In this category the true epiphytes are considered. True epiphytes are herbs and woody plants that germinate on other woody plants (Benzing 1995) (for hemiepiphytes, see climbers). Diversity of ground-rooted organisms increases with sampled area (MacArthur & Wilson 1967; Connor & McCoy 1979), whereas epiphytic diversity increases with number and diversity of sampled host-trees. Epiphytes are especially known for their ability to endure stress, meaning long periods of drought and high irradiation (*e.g.*, Benzing 1989, 1990) and are slow growing due to low water and nutrient availability. Epiphytic species richness is therefore not expected to show any change related to logging on the time scale considered in this research. Nevertheless, to complete the whole range of plant species in Greenheart forest, we included an enumeration of the epiphytic vegetation during the logging phase. Comparison of the vertical distribution and diversity patterns of vascular epiphytes with those patterns for three other forest types in the Guianas are treated elsewhere (Ek *et al.*, submitted, chapter 3).

Plot design and data collection

The one-ha plots were enumerated with the following intensity:

- canopy trees (dbh≥20 cm), large woody lianas (dbh≥10 cm), palms (height≥2 m), and shrubs (height≥2 m) in each 100 x 100 m plot, divided in 20 x 20 m subplots;
- small canopy trees and understorey trees (5 cm≤dbh<20 cm), lianas (height≥2 m), palms (1m≤height<2 m), shrubs (50cm≤height<2 m), and herbs (50 cm≤height<2 m) in 10 x 10 m subplots, one for each 20 x 20 m subplot;
- saplings of canopy and understorey trees (2 cm≤dbh<5 cm), lianas (50 cm≤ height<2 m), palms (height<1m), shrubs (height<50 cm), and herbs (height<50 cm) in 5 x 5 m subplots, one for each 10 x 10 m subplot;
- seedlings of canopy and understorey trees (height>1.5 m and dbh< 2 cm), lianas (height<50 cm) in 2 x 2 m subplots, one for each 5 x 5 m subplot;

More details on plot-lay out and sampling method can be found in Alder & Synnott (1992), and Van der Hout (1996).

For all 24 one-ha plots all lianas and trees were enumerated. Herbs and shrubs were enumerated for all Pibiri and Waraputa plots, so 18 one-ha plots in total. In the other six plots, however, only presence of species for each subplot were recorded.

For each sampled ground-rooted individual, species name, coordinates within the plot, height, and diameter were recorded. For the trees the height was measured using a clinometer, the diameter with a diameter-tape or caliper. For lianas, herb, shrubs, and small palms height up to 2 m was measured using a measuring tape on a pole of

4 m high. Further heights were estimated using the length of the pole as reference. For the group of the lianas, the diameter of the host and climbing method was recorded. For lianas, members of a clonal group with above-ground connections were regarded as one individual. Individuals growing in the plot, but rooting outside, where not included. Position in the plot was measured from the point were an individual rooted.

Vascular epiphytes were enumerated during logging (March - July 1994). A total of 105 trees was sampled, representing all harvested tree species. Trees were selected *a priori* based on the felling lists. For epiphytes the species of the host-tree, height zone on the tree, relative occurrence within the height-zone, number of individuals, and substrate were recorded.

For all species vegetative voucher collections were made and deposited in Utrecht (U). Position of individuals in the plot was recorded using two compass bearings from fixed corner markers in the plots (20 x 20 m grid) for the large individuals, and from the temporary marker (10 x 10 m grid) for the smaller individuals.

In the exploited forest sites, the size and position of gaps and skid trails were measured. Gap area, measured as a vertical projection of a canopy opening enclosed by the crown edges of the surrounding trees (Putz 1984a), was estimated using an octagon method connecting eight gap edge points at 45° intervals (after Lawton & Putz 1988; Jans *et al.* 1993). Irregular shaped gap edges between the 45° intervals were recorded when appropriate. Afterwards, we defined an extended borderzone, an uncertainty zone of 0.5 m inwards and outwards of the measured gap border (the 'gap edge'), and different gap zones in and around gaps using a GIS-system (PC-Raster, Van Deursen & Wesseling 1993). Skid trails were traced using signs of low secondary vegetation, damage to remaining large trees, and the presence of stumps of harvested trees. The skid trail area was estimated by measuring the position of trail edges in relation to the 20 x 20 m grid of the plots, combined with positions of points where the trails divided in two or more trails, or changed direction. Those points were measured using two compass bearings from corner markers in the plots. An extended borderzone of 0.5 m inwards and outwards was calculated afterwards. Lastly, in relation to the 20 x 20 m grid of the 9 one-ha plots in the exploited sites, the inclination of the terrain in NS and EW direction was measured using a clinometer, and related to a nearby creek as zero-level.

Data analyses

General

For the exploited forest sites, the GIS system was used to integrate the measurements of the species composition, their relative abundance and location, and the location, size and form of gaps and skid trails. The forest was classified in a 'damage zonation' (Fig. 2) in which we distinguished between gaps, large gaps (≥ 300 m^2), small gaps (< 300 m^2), skid trails, overlapping gap and skid trail, outer gap (a zone up to 15 m from gap edge) and forest interior (further than 15 m from a gap edge). For the outer gap zone we used a distance of 15 m from the gap edge, due to

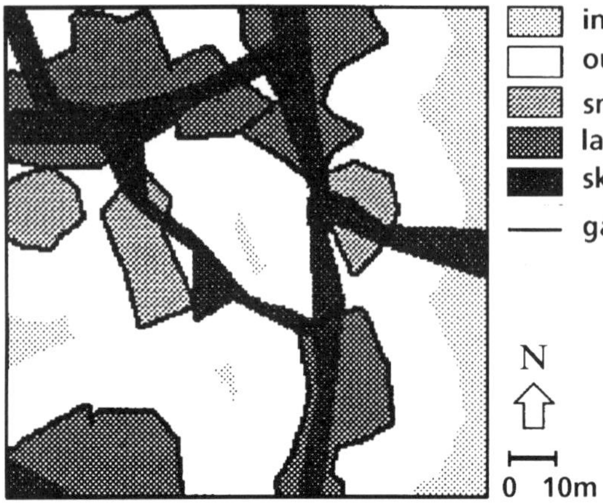

Figure 4.2
Damage zonation.

edge). For the outer gap zone we used a distance of 15 m from the gap edge, due to the fact that we calculated* significantly increased light levels up to a distance of 0.5 x canopy height. Greenheart forest had a general height of 30-35 m.

* Under the assumption that light levels around gaps were influenced by the size and orientation of a gap towards the solar track, we calculated for all points in and around gaps a yearly averaged time of exposure to direct light through the gap as a percentage of full direct sunlight. The attenuation of light with distance from a gap was considered dependent on the height of the surrounding canopy and the density of surrounding vegetation. The increased light levels over the plot after gap formation were scaled logarithmically. For equitations used see appendix 6.1.

Pattern analyses

Due to the sampling method abundances and frequencies of species are evaluated separately for the different subplot areas; for canopy trees and large woody lianas over the whole one-ha, for smaller trees and lianas on the basis of 10 x 10 m subplots. Abundances of species in relation to damage zones (*e.g.,* skid trails, large gaps, small gaps) are evaluated per unit area (m^2). The latter analyses were performed for the exploited forest sites. In those analyses we used the size classes of individuals of which we assumed that they established shortly after logging; for the trees the size class of individuals with a diameter between 1 and 5 cm, and for the lianas the individuals with a height of more than 2 meters.

To evaluate the homogeneity of the different plots, plots were pooled in an observed cumulative species curve and a mean random cumulative species curve. The observed curve was obtained by pooling the plots in order of enumeration and the mean random curve was obtained by randomly pooling the plots twenty times.

The multivariate analyses we performed on the total data-set were mainly clustering using BIOPAT (Hogeweg & Hesper 1972, 1981), correlation coefficients (Pearson product moment) and multiple linear regression. These methods were used to compare the patterns of similarities and differences among plots, subplots, and groups of species. Species used in the calculations are those of which the level of occurred, in which several different species could not be distinguished in the field using vegetative characteristics. Those groups were left out of the analyses, or counted as one species for total number of species.

(Dis)similarity matrices were calculated with either absence/presence or normalised data, and mean square distances. Agglomerative clustering was performed using Ward's averaging (minimum increase of the mean sum of squares of distances within clusters (Ward 1963)). Due to the tendency of Ward's averaging to form clusters with an equal number of members, Group averaging was used as a control. Normalisation indicates that the abundances of species in each sample were standardized to their proportions of the total sample abundance. This approach is specifically appropriate for species-rich tropical data-sets (Newbery 1991). Kruskal Wallis one-way analysis of variance was used to test differences between groups, and to determine the importance of cluster members. Similarity between plots was also calculated using Jaccard's Index (1912). The Jaccard Index gives the proportion of species out of the total species list for two plots or subplots, which is common to both. The Wilcoxon matched pair test was used for pair-wise comparison of the species composition for different plots and forest zones. This non-parametric procedure weights differences in rank order between species and also uses rank ordering of the differences between species.

Patchiness of the vegetation was evaluated using an autocorrelation between dissimilarity in species composition of subplots and their distance. Dissimilarity of subplots was based on species composition (mean square distances) and the distance between subplots was expressed as number of circles of neighboring plots (1-4).

Number of rare species can be defined on different levels. We distinguished between species that occurred only at one site, only in one plot, only in one subplot, or only with one individual. The intensity of sampling, however, influences the chance of a species to be labelled 'rare'. In order to evaluate differences in number of rare species between Pibiri and the exploited sites at equal sampling intensities, we randomly selected plots out of Pibiri. To compare with the unlogged referential plots, we randomly selected three plots, one for each of the three kilometer squares (Fig. 3) in which the plots were situated. This procedure was repeated 10 times. To compare with the exploited plots we randomly selected six plots. The same procedure was applied for the comparison of logged and unlogged plots of the exploited sites.

Multiple linear regressions were used to test the relationship between species richness after logging, and the extent of logging damage. Species richness was expressed as number of species per plot after logging, number of species before logging (represented by number of species in the unlogged referential plot), and number of species only present after logging (newly established species). For the exploited sites there are no data of the situation before logging. For these plots we used the number of species of the unlogged plot as representative for that pre-logging situation. Logging damage was expressed as number of harvested trees, total gap area, total skid trail area, total combined gap and skid trail area, and total amount of remaining forest (interior). None of these variables showed direct correlations with

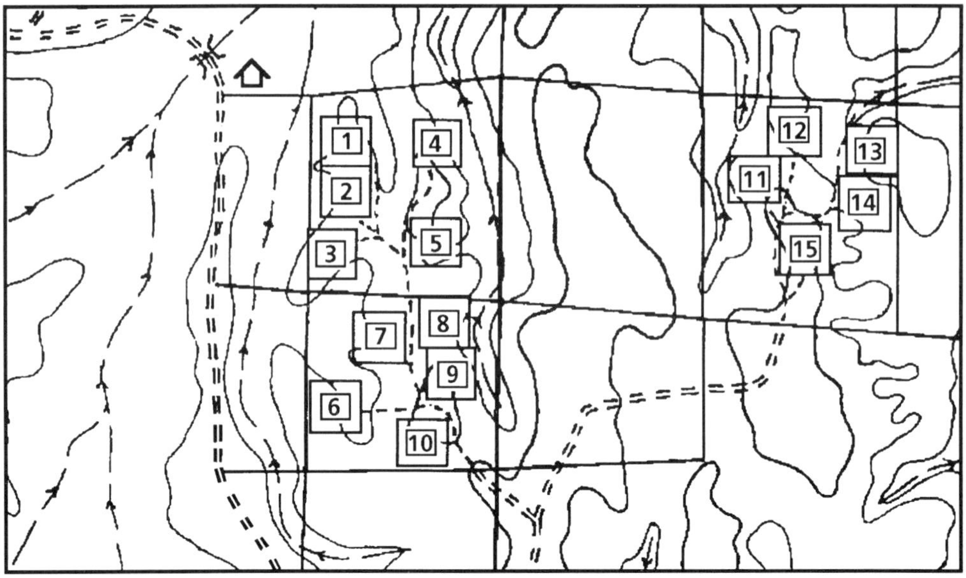

Figure 4.3
Location of the 15 one-ha plot in the West-Pibiri Compartment, Mabura Hill, Guyana.

each other and were therefore used separately as parameters for logging damage. A stepwise forward method was used, and a linear regression is only given when the result of the regression was significant with three or less variables.

For all the statistical analyses Statistica 4.5 (StatSoft 1993) was used.

Results

Botanical diversity of unexploited Greenheart forest

In 15 one-ha plots in unexploited Greenheart forest a total of 405 species were found: 172 tree species, 132 liana species, 36 undergrowth species, and 65 epiphytic species. A list of all species and their authorities is given in appendix 1. The number of species ranged between 125 and 170 species per ha. The most abundant species (> 2% of all individuals) are listed in table 1.

1 Trees

Greenheart (*Chlorocardium rodiei*), contributed the highest percentage of individuals of all 118 canopy tree species found, and was present in all plots. Wirimiri Kakaralli (*Lecythis confertiflora*) and Greenheart contributed more than 10 percent of all individuals. Of the smaller trees, Waiaballi (*Tapura guianensis*), an understorey tree, presented more than 10% of all individuals over 15 hectare. In the size category of smaller trees (with dbh between 5 and 20 cm) also many small individuals of future canopy trees were found. Other frequent 'true' understorey trees are: black Karishiri (*Oxandra asbeckii*) and Adebero (*Papayrola guianensis*).

Table 4.2
Mean number of epiphytes per investigated tree in Greenheart forest, near Mabura Hill, Guyana, both Greenheart trees and other host tree species. Ntot= total number investigated trees, Nep= number of trees carrying epiphytes, ***: p< 0.001, ns= not significant.

Plots	Ntrees		Greenheart				Others				Sign.
	Ntot	Nep	N	Nep	mean	stdev	N	Nep	mean	stdev	
All	105	73	57	53	4.94	3.47	48	20	2.16	0.85	***
'8'	15	15	9	9	7.22	4.60	6	6	4.83	0.75	ns

2 Lianas

Within the group of large woody lianas (with a dbh >10 cm) 32 species were found. Most individuals (24.4%) are contributed by one species of Devildoer (*Anomospermum grandifolium*) occurring in 13 of the 15 plots. No large woody liana appeared in all plots, and 9 of the 15 (60%) do not belong to the abundant species for the smaller size classes. Smaller lianas were more evenly distributed over the plots, and 17 species occurred in all plots. Of those 17 species, Brown Belly (*Connarus perrottetii* var. *rufus*), Supple Jack (*Rourea pubescens* var. *spadicea*), Nibi (*Heteropsis flexuosa*), and Monkey syrup (*Maripa scandens*) contributed more than 5% of all individuals.

3 Undergrowth

Palms play an important role in the undergrowth of Greenheart forest. One small palm species, *Bactris oligoclada,* contributed 43.5% of all measured individuals over all 15 one-ha plots. The second most abundant species was also a palm, *Bactris balanophora,* with 10.9 %, followed by a kind of Mucru (*Ischnosiphon puberulus* var. *scaber*). In this growth form group also the saprophyte *Voyria corymbosa* was found. This species, however, shows a temporal pattern, *i.e.* it is only present during the wetter periods of the year. *Voyria corymbosa* and other saprophytes could therefore be much more abundant than listed in table 1.

4 Epiphytes

Epiphytes were sampled on 105 trees. Of these trees, 57 were Greenheart trees and 48 were other host-tree species (table 2), such as *e.g.*, Purpleheart (*Peltogyne venosa*), Baromalli (*Catostemma* spp.), Morabukea (*Mora gonggrijpii*) and Crabwood (*Carapa guianensis*). Only 73 trees carried epiphytes. *Codonanthe crassifolia* and *Maxillaria uncata* were the most frequently found epiphytes. More Greenheart trees (93%) carried epiphytes than non-Greenheart trees (42%), and Greenheart carried significantly (p<0.001) more epiphytic species per tree (mean=4.94 species per tree, stdev=3.47) (table 2). Only in one plot, plot 8, all investigated trees carried epiphytes, and there Greenheart trees carried comparable numbers of epiphytes as the other host species (mean 4.83, stdev= 0.75). The trees in this plot contained many ant nests (ant

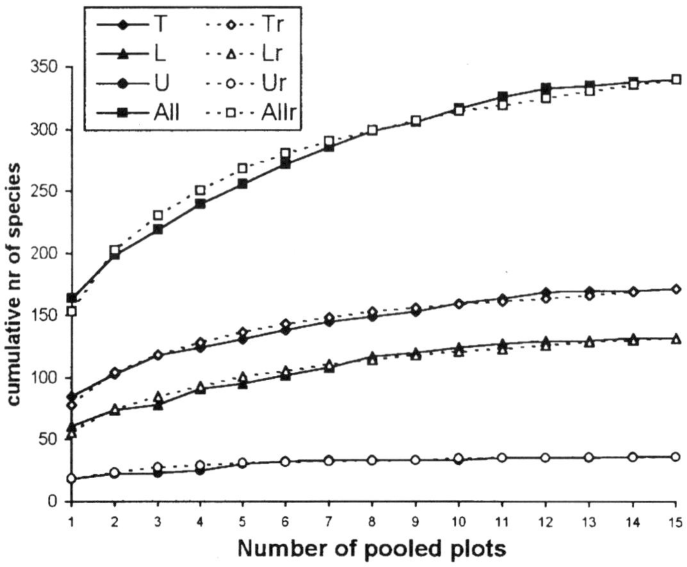

Figure 4.4
Observed and random cumulative species curves for 15 one-ha plots of Greenheart dominated forest in the West-Pibiri compartment, near Mabura Hill, Guyana. T= trees, L= lianas, U= undergrowth, All= all growth forms, r= random.

gardens). Ant gardens are among the most abundant and distinctive ant-plant associations in Neotropical forest canopies and the ant nests are associated with a particular set of specialised plants (Tobin 1995; Benzing 1995). The epiphytes appear to depend on the ants both for the nutrients with which ants enrich the substrate and for seed dispersal (Longino 1986; Benzing 1995). Specific epiphytes, found only on these ant gardens were *Peperomia macrostachya* (Piperaceae), *Codonanthe calcarata* (Gesneriaceae), *Gravisia* sp. and *Streptocalyx longifolius* (Bromeliaceae). Other epiphytes, such as *Anthurium gracile* (Araceae) and *Codonanthe crassifolia* (Gesneriaceae) were found on the ant nests, but not exclusively.

All plots were pooled in an sequential observed cumulative species curve and a mean random cumulative species curve, for all species together and per growth form. The observed curve (Fig. 4) hardly deviated from the random curve. This result suggests a well mixed forest, although there is no complete mixing. The plots 3-5 seem to harbour relatively less new species than plots 11-12. The species of different growth forms, however, showed very unequal abundances, as shown in table 1 and figure 5. In figure 5 the cumulative percentage of abundances of the species ranked 1-10 per growth form are given. Undergrowth species showed the highest curve, indicating that this growth forms had the highest number of dominant species.

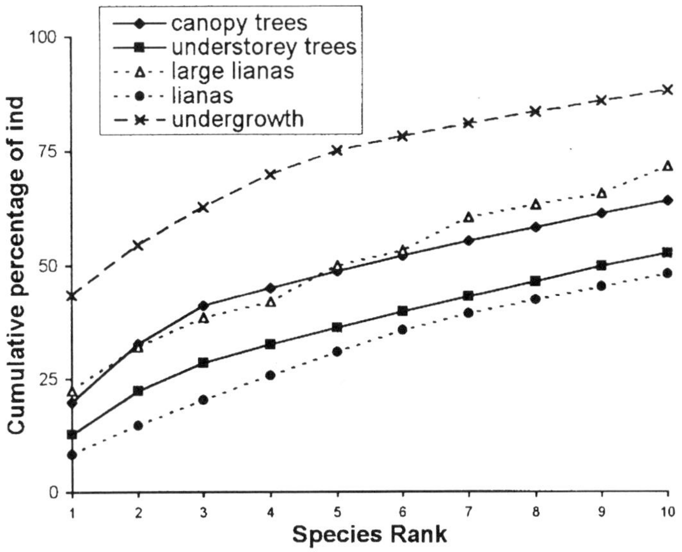

Figure 4.5
Cumulative percentage of individuals for the ten highest ranked species, for five growth forms. Species are ranked according to their abundance.

Botanical diversity of unexploited and exploited Greenheart forest

Species richness

In comparison with the 15 ha of unexploited Greenheart forest, we found an additional 38 tree species, 17 liana species, and 1 undergrowth species number in the unexploited referential plots from the other sites (Waraputa, Reserve, and '2km'). Including also the exploited plots added another 41 tree species, 27 liana species and 2 undergrowth species. In total, over the 24 one-ha plots a number of 466 botanical species were found (appendix 1, table 3). The number of species per one-ha plot ranged between 120 and 189 for the exploited plots (table 3). For all unexploited plots together this number ranged between 122 and 179. For all plots, both exploited and unexploited, the trees were the most species-rich growth form, followed by the lianas and by the undergrowth species. The sequential observed cumulative species curve and a randomized cumulative species curve for all unexploited plots (N=18) is given in figure 6. Both the sequential and randomized curve showed that, except for the undergrowth species, the Pibiri site and the other sites cannot be considered as homogeneous. For the lianas and trees the observed curves started to level off for the Pibiri plots (1-15), and for those two growth forms adding additional sites increased the level of species richness considerably. For the undergrowth species, already with 11 plots 90% of all undergrowth species were found.

High species richness can be found both in unexploited and in exploited plots

Table 4.3
Species richness per growth form for all 24 one-ha plots in Greenheart dominated forest near Mabura Hill, Guyana. The number of rare species is given, where rareness is considered on the level of sites, of plots, subplots, and on the level of occurring only with one individual.

	Total # sp	# sp			# rare species			
	Plot	Trees	Lianas	Under-growth	Site	Plot	Subplot	Ind
Pib 1	163	85	61	17	104	1	0	0
Pib 2	142	75	56	11		3	1	1
Pib 3	142	80	46	16		2	1	1
Pib 4	150	75	61	14		2	2	1
Pib 5	148	80	49	19		4	3	3
Pib 6	147	81	50	16		2	1	1
Pib 7	152	79	54	19		5	3	1
Pib 8	160	79	67	14		3	3	3
Pib 9	141	71	54	16		4	4	3
Pib 10	162	83	57	22		3	2	2
Pib 11	148	80	53	15		4	3	1
Pib 12	170	98	58	14		3	2	1
Pib 13	125	65	39	21		1	1	1
Pib 14	148	74	54	20		3	2	2
Pib 15	132	70	43	19		1	1	1
War 2	122	77	39	6	18	5	3	2
War 1*	120	57	47	16		3	3	1
War 3*	143	81	46	16		3	2	1
Res 1	133	72	43	18	25	5	5	4
Res 2*	165	88	62	15		5	4	1
Res 3*	167	94	60	13		6	5	2
2 km 3	179	96	69	14	27	5	4	1
2 km 1*	179	99	59	21		7	4	1
2km 2*	189	100	68	21		6	5	3
Total	**466**	**251**	**176**	**39**	**174**	**86**	**64**	**38**
Pibiri (N=15)								
Total N of sp	**340**	**172**	**132**	**36**				
perc. of sp	**100**	**50.6**	**38.8**	**10.6**				
Mean N of sp / plot	**148.5**	**78.3**	**53.0**	**16.7**				
Natural forest (N=18)								
Total # sp	**405**	**220**	**149**	**36**				
% sp	**100**	**54.3**	**36.8**	**8.9**				
Mean #sp/plot	**148.0**	**78.8**	**52.6**	**16.1**				
Logged forest (N=6)								
Total # sp	**352**	**178**	**147**	**27**				
% sp	**100**	**50.6**	**41.8**	**7.7**				
Mean #sp/plot	**160.2**	**86.3**	**57.2**	**17.0**				

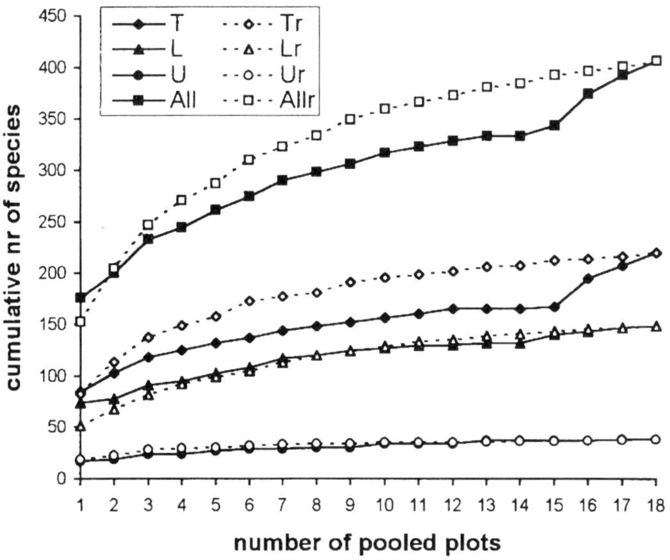

Figure 4.6
Observed and random cumulative species curves for 24 one-ha plots of Greenheart dominated forest, near Mabura Hill, Guyana. T= trees, L= lianas, U= undergrowth, All= all growth forms, r= random.

(table 3). Mean number of species per plot was higher in exploited plots for all growth forms. Lianas showed the highest increase in percentage of total number of species and thus the relative percentage of liana species after logging was higher.

Number of rare species on the level of sites is not significantly different for the three sites with three plots (data not shown) and the difference between those three sites with the Pibiri site is mainly influenced by the larger number of sampling plots at Pibiri. There was a significant difference (ANOVA, $p<0.001$) in mean number of rare species per plot and per subplot between randomly selected unexploited plots of the Pibiri site and the unexploited referential plots of all other sites (table 4). The three referential plots from the exploited sites contained more rare species. The same significant difference in number of rare species per plot and per subplot occurred between the exploited plots and randomly selected Pibiri plots. The exploited plots additionally showed a significant higher number of species that occurred with only one individual. In the exploited sites, no significant differences between mean numbers of rare species were found between exploited and unexploited plots.

The correlation between the number of species in the different growth forms for unexploited and exploited Greenheart forest based on the species richness of subplots (20 x 20 m), is given in table 5. Species richness of lianas is positively correlated with total species diversity and species diversity of trees, both for unexploited and exploited subplots. Species diversity of undergrowth species, however, is only slightly correlated with total species diversity, and only for unexploited plots.

Table 4.4
Analyses of variance (ANOVA) of number of rare species occurring in Pibiri and the exploited sites, and between exploited and unexploited plots. Rareness is defined as occurring only in one plot, only in one subplot, and only with one individual.

	Pibiri			Others unlogged			F value	sign
Rare:	N*	Mean	sd	N	Mean	sd		
only in 1 plot	3	15.90	4.95	3	22.73	5.53	25.42	***
only in 1 subplot	3	11.26	3.69	3	18.26	4.54	42.81	***
only with 1 ind	3	10.56	4.45	3	12.23	3.34	2.69	ns
	Pibiri			**Others logged**			**F value**	**sign**
Rare:	N*	Mean	sd	N	Mean	sd		
only in 1 plot	6	6.18	2.16	6	8.19	2.27	15.52	***
only in 1 subplot	6	3.20	2.49	6	7.22	4.14	29.06	***
only with 1 ind	6	2.56	2.22	6	6.03	1.90	9.31	**
	Others unlogged			**Others logged**			**F value**	**sign**
Rare:	N	Mean	sd	N*	Mean	sd		
only in 1 plot	3	16.43	5.94	3	18.33	9.38	0.87	ns
only in 1 subplot	3	13.70	4.04	3	13.00	4.68	0.53	ns
only with 1 ind	3	10.93	3.71	3	11.73	5.56	0.43	ns

N*: Number of used plots for the comparison. The means were calculated using 10 random draws of this number of plots from the set, and calculating the number of rare species using the total combination of plots.

Similarity in species composition between sites, plots, and within plots

The Jaccard Index (table 6) was calculated to compare the similarity in species composition between the plots. Plots from one site were more similar in species composition to each other, compared to plots from other sites. In Pibiri (plot 1-15) highest similarity occurred always with another plot from Pibiri, and mostly with plots in closest vicinity (table 6, Fig. 2). This was also true for the exploited sites, in which the similarity between the exploited plots for each site showed the highest values. Similarities between sites occurred mainly through unexploited plots. For instance, highest similarity between the sites was always with the unexploited plot of 2km (2km UL). This plot has a Jaccard Index of 0.43 with the unexploited plot of Reserve (R UL), of 0.44 with one of the exploited plots of Waraputa (log 2), and of 0.40 with one of the Pibiri plots (plot8).

The Jaccard Index was calculated with presence/absence data. Agglomerative cluster analyses were used to calculate similarities between plots based on the relative

Table 4.5
Pearson product moment correlations between the number of species in the different growth forms at the level of one-ha plots for unexploited and exploited Greenheart forest near Mabura Hill, Guyana. Values indicated **bold** are significant, *: p<0.05, **: p<0.01.

	Trees		Lianas		Undergrowth	
	Natural	Logged	Natural	Logged	Natural	Logged
Species	**0.87** **	**0.96** **	**0.86** **	**0.93** **	**0.58** *	0.42
Trees	-	-	**0.60** *	**0.82** *	0.35	0.24
Lianas			-	-	0.37	0.37
Undergrowth					-	-

abundance of species. The resulting similarities can be visualized using dendrograms. In Fig. 7 the similarities between all 24 plots are given, for all species and growth forms. The various sites clustered together in almost all dendrograms (except the one for the herbs and shrubs), showing a strong influence of geographical variability. For the undergrowth species the two unexploited plots of Waraputa and Reserve clustered together with the unexploited Pibiri plots, and only one of the Waraputa plots (WLOG1) seemed to be separated from all others. This plot contained as only exploited plot the two new species that were found in the exploited plots. For the trees, the plots of '2km' were more similar to the Pibiri plots than to the two sites that are closer to Mabura Hill. For the lianas, only the unexploited plot from the Reserve site deviated from a strict site specific clustering.

These results show that geographical distance plays a role in determining similarities in species composition between plots. To investigate a possible distance dependence within the one-ha plots, we calculated the Pearson's product moment correlation between (dis)similarity of the subplots based on species composition and their distance to each other (table 7). For the lianas, most plots showed a significantly positive correlation. For the trees, however, this correlation was found in less plots. Distance explained a larger part of the variance between subplots for the species composition of lianas, compared to trees. Thus, lianas have a more patchy distribution on the scale investigated.

Logging damage

The previous analyses for unexploited and exploited Greenheart forest were performed with all 24 one-ha plots. To investigate the influence of logging damage on the species composition of plots, we used the plots from the exploited sites (Waraputa, Reserve, and 2km).

The difference in species composition between plots within one site was tested using a Wilcoxon matched pair test. The results for the exploited sites are given in table 8. For the two sites logged approximately 6 years ago (Waraputa and Reserve), the unexploited plots showed a significant different species composition from the two

Table 4.6
Matrix of similarity (Jaccard Index) of 24 one-ha plots of Greenheart forest, near Mabura Hill, Guyana.

	Pibiri											
	1	2	3	4	5	6	7	8	9	10	11	12
1												
2	*0.58*											
3	0.55	*0.56*										
4	0.55	0.51	0.52									
5	0.52	0.54	0.48	*0.56*								
6	0.48	0.48	0.45	0.47	*0.55*							
7	0.50	0.53	0.45	0.52	0.53	*0.54*						
8	0.51	0.51	0.47	0.48	0.54	0.54	0.50					
9	0.46	0.44	0.45	0.49	0.49	0.52	0.52	*0.56*				
10	0.51	0.51	0.48	0.51	0.53	0.51	*0.54*	0.54	0.55			
11	0.44	0.44	0.49	0.49	0.47	0.50	0.51	0.51	0.50	0.55		
12	0.47	0.47	0.46	0.40	0.48	0.51	0.47	0.53	0.47	*0.56*	*0.56*	
13	0.44	0.49	0.46	0.45	0.47	0.49	0.45	0.48	0.50	0.51	*0.58*	0.48
14	0.50	0.47	0.50	0.50	0.51	0.53	0.48	0.54	0.52	0.55	0.54	0.50
15	0.45	0.45	0.54	0.49	0.47	0.48	0.44	0.48	0.49	0.47	0.54	0.50
W UL	0.28	0.28	0.26	0.26	0.26	0.28	0.25	0.31	0.27	0.27	0.26	0.27
W log1	0.25	0.26	0.23	0.25	0.28	0.29	0.26	0.28	0.28	0.28	0.29	0.29
W log2	0.28	0.27	0.27	0.27	0.27	0.26	0.24	0.29	0.26	0.29	0.29	0.30
R UL	0.31	0.31	0.33	0.32	0.31	0.35	0.33	0.32	0.32	0.32	0.31	
R log1	0.29	0.28	0.27	0.28	0.27	0.29	0.28	0.31	0.28	0.26	0.28	0.30
R log2	0.29	0.29	0.28	0.31	0.27	0.29	0.26	0.30	0.29	0.28	0.31	0.31
2km UL	0.38	0.36	0.38	0.36	0.35	0.36	0.36	**0.40**	0.35	0.36	0.37	0.35
2km log1	0.32	0.31	0.34	0.33	0.29	0.31	0.30	0.36	0.32	0.33	0.34	0.33
2km log2	0.32	0.29	0.27	0.30	0.28	0.28	0.32	0.33	0.30	0.29	0.32	0.30

exploited plots. The two exploited plots were not significantly different. Ten years after logging, at 2km, no significant difference in species composition was found on the level of one-ha plots.

Correlations (Pearson's product moment) between species richness per plot and logging damage per plot are given in table 9. The total number of species after logging was significantly positive correlated with number of species before logging, but did not show direct correlations with logging damage. Only the new liana species after logging and the number of new undergrowth species after logging showed correlations with the number of dead standing trees. Between logging damage classes there was a negative correlation between the extent of large gap area and the amount of remaining interior. Time after logging was correlated with almost all species richness variables, before and after logging. This correlation, however, is meanly due

Table 4.6 continued; UL= unlogged, LOG= logged. *Italic* values are highest similarity value for each plot. **Bold** values are mentioned in the text.

	Pibiri			Waraputa			Reserve			2km		
	13	14	15	UL	LOG1	LOG2	UL	LOG1	LOG2	UL	LOG1	LOG2
1												
2												
3												
4												
5												
6												
7												
8												
9												
10												
11												
12												
13												
14	0.54											
15	0.51	*0.58*										
W UL	0.28	0.30	0.28									
W log1	0.25	0.30	0.25	0.45								
W log2	0.27	0.27	0.25	*0.49*	*0.55*							
R UL	0.30	0.30	0.28	0.37	0.39	0.38						
R log1	0.25	0.29	0.24	0.36	0.38	0.39	0.34					
R log2	0.28	0.30	0.28	0.35	0.36	0.41	0.36	*0.63*				
2km UL	0.36	0.39	0.32	0.42	0.38	**0.44**	***0.43***	0.39	0.42			
2km log1	0.32	0.32	0.31	0.35	0.33	0.38	0.42	0.35	0.41	*0.51*		
2km log2	0.28	0.29	0.31	0.29	0.28	0.30	0.39	0.35	0.38	0.45	*0.52*	

to the fact that the 2km site, the site that had been logged 10 years before enumeration, was also the richest site. The two other exploited sites were both logged approximately six years before the enumeration. We concluded that time after logging, therefore, could not be further used in the analyses.

Multiple linear regressions of species richness after logging and the amount of logging damage for the same six plots, showed again that the total species richness after logging depended mainly on the species richness before logging (table 10). Logging damage, and in particular the total extent of gap area and large gap area, was important for the establishment of new, invading species after logging. It was not possible to obtain a significant multiple regression model for the total number of tree species after logging including less than four variables. The number of new tree species after logging, and the number of new liana species after logging, were

Figure 4.7
Similarity of 24 one-ha plots in Greenheart dominated forest near Mabura Hill, Guyana. UL= unlogged, LOG1 & LOG2= logged, R= Reserve, W= Waraputa, KM2= 2km, P= Pibiri.

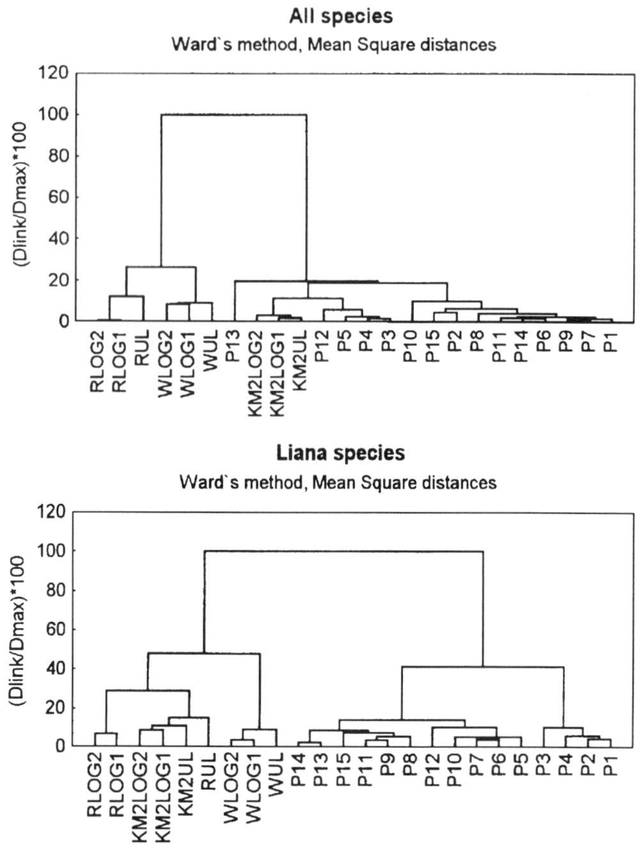

negatively correlated with the number of dead standing trees and the total extent of gap area, while positively correlated with total amount of overlapping gap and skid trail area. The species that were not present in the unexploited plots but were only found in the exploited ones could be found in the size class of trees between 1 and 5 cm dbh, with a height class of more than 2 meters for the lianas, and in all size classes for the undergrowth species.

The results for the multiple regression were obtained at plot level, for six exploited plots in the exploited sites. For each exploited plot the number of species of the unexploited plot of the same site was regarded as representative for its previous unexploited situation. On the level of subplots, this assumption can no longer be made. Therefore, the correlation between species numbers, abundance, and logging damage on subplot level could only be evaluated using the spatial effects per subplot. In this analysis only the trees and lianas were used and the results are listed in table 11. Correlations were calculated between the total species richness, and abundance, per size class, and the amount of area occupied by a damage class for each subplot. A positive correlation between, *e.g.*, the number of liana species in the 10 x 10 m

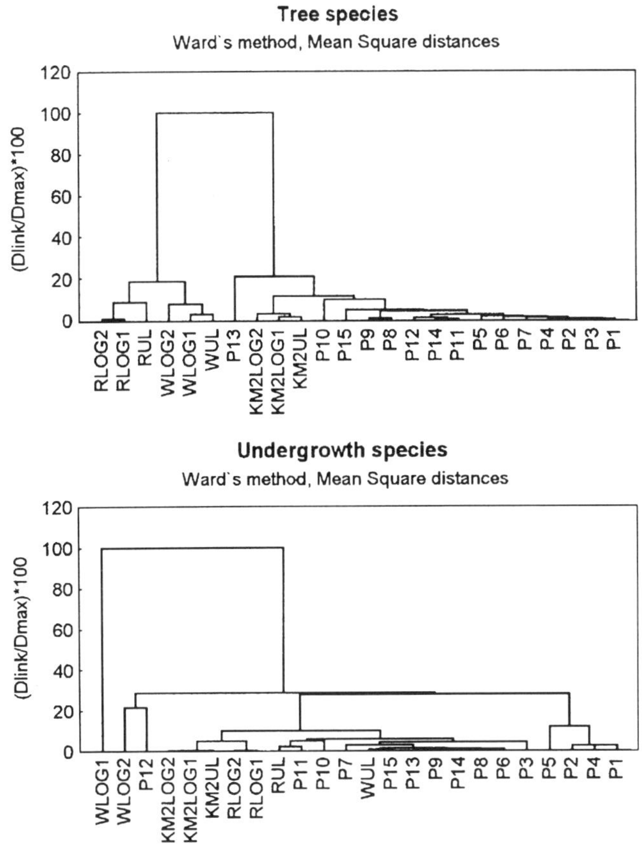

subplots and the amount of skid trail, would indicate that with increased amount of skid trail area, there were increased numbers of liana species in this size subplot. Lianas showed more significant correlations between logging damage and number of species or abundance than trees.

Large gaps versus small gaps

The influence of gap size on the species composition of plots was evaluated using the plots from the exploited sites (Waraputa, Reserve, and 2km). For these nine plots the position and size of gaps were available, and gaps were divided in small (<300 m^2) and large (≥ 300 m^2) gaps. For each plot, the number of individuals per square meter in small gaps and in large gaps was calculated, for each liana and tree species. Although hampered by the absence of small gaps in the unexploited plot of the Reserve site, this analysis showed that for five out of the eight remaining plots there existed a significant difference in species composition between small and large gaps (table 12). For all sites there was a significant difference between the species composition in large gaps in unexploited plots, compared to the species composition

Table 4.7
Correlation between (dis)similarity of subplots based on their species composition and their distance. Correlation calculated as Pearson's product moment, dissimilarity calculated with mean square distances, distance expressed as number of circles of neighboring plots (1-4).

Site	Plot	Corr. Coef. Lianas (mnsq)	Corr. Coef. Trees (mnsq)
Pibiri	1	-0.02	0.02
	2	**0.18** **	0.05
	3	**0.11**	**0.21** ***
	4	**0.17** **	0.00
	5	**0.19** **	**0.25** ***
	6	**0.12** *	**0.15** **
	7	**0.17** **	0.10
	8	-0.09	-0.02
	9	**0.14**	**0.35** ***
	10	-0.09	0.09
	11	**0.23** ***	0.09
	12	**0.13** *	**0.12** *
	13	**0.12** *	0.01
	14	0.07	**0.30** ***
	15	**0.19** **	**0.23** ***
Waraputa	Unlogged	**0.13** *	0.11
	Logged1	**0.20** ***	0.11
	Logged2	**0.12** *	0.10
Reserve	Unlogged	-0.05	0.03
	Logged1	**0.19** **	**0.12** *
	Logged2	**0.18** **	0.10
2 km	Unlogged	**0.12** *	0.00
	Logged1	**0.17** **	**0.24** ***
	Logged2	**0.13** *	0.00

(Dis)similarity calculated with mnsq and Jaccard, N=300. Correlation calculated with Pearson's product moment. Values indicated **bold** are significant: *: p<0.05, **: p<0.01, ***: p< 0.001.

in large gaps in the exploited plots. For the difference in species composition of small gaps between unexploited and exploited plots, half of the remaining four plots showed significant differences. As overlapping gap and skid trail area is important for the establishment of new species of trees and lianas (table 10), we investigated the

Table 4.8
Wilcoxon matched pair test for differences in species composition between unexploited and exploited plots from the same site, near Mabura Hill, Guyana.

		Valid N	T	Z	p-level
Waraputa	Unlogged & LOG1	466	5609.50	2.69	**
	Unlogged & LOG2	466	4082.00	3.40	***
	LOG1 & LOG2	466	6755.00	0.28	ns
Reserve	Unlogged & LOG1	466	8910.00	2.25	*
	Unlogged & LOG2	466	9113.00	2.53	*
	LOG1 & LOG2	466	6202.50	1.54	ns
2 km	Unlogged & LOG1	466	10850.50	0.83	ns
	Unlogged & LOG2	466	11899.00	0.16	ns
	LOG1 & LOG2	466	8314.50	1.57	ns

influence of combined skid trail and gap area on the difference in species composition between unexploited and exploited plots for gaps of the same size, by excluding the individuals growing on combined gap and skid trail area from the analysis. For the 2km site, 10 years after logging, the difference between the species composition of both small gaps and large gaps were no longer significant between the unexploited one-ha plot and one of the exploited plots. All other species compositions of gaps of the same sizes of unexploited and exploited plots remained significantly different.

Behaviour of individual species

Lastly, we wanted to know if we could find specific species that correlated with logging damage. In table 13 the correlation of individual species with different logging damage classes is given. All species that are listed showed a significant correlation with one of the damage classes, with Greenheart as the only exception.

Discussion and conclusions

Alpha diversity of unexploited Greenheart forest

The comparison with other studies regarding species richness of unflooded tropical rain forest is hampered by different sampling methods, number of included growth forms and size limits. Most studies list only trees (and sometimes lianas) with a dbh > 10 cm (*e.g.*, Campbell *et al.* 1986; Gentry 1990). Studies that investigate all vascular plant growth forms are hardly undertaken. We can use our results of trees and lianas with a dbh > 10 cm for a comparison. In the unexploited Greenheart forest in Pibiri we found 405 botanical species, with a mean of 41.5 tree species and 8.5 liana species

Table 4.9
Pearson's product moment correlation between species richness and logging damage for 6 exploited plots in Greenheart forest, near Mabura Hill, Guyana.

	TIME	All species			Trees			Lianas		
		sp	N of old	new	sp	N of old	new	sp	N of old	new
SP	**0.72**									
SPOLD	***0.98***	**0.82**								
SPNEW	0.01	0.60	0.17							
TRE_SP	0.63	***0.97***	0.73	0.65						
TRE_OLD	***0.99***	0.64	***0.95***	-0.10	0.55					
TRE_NEW	0.16	0.78	0.33	0.80	0.80	0.05				
LIA_SP	**0.58**	***0.91***	0.70	0.62	**0.82**	0.48	**0.85**			
LIA_OLD	***1.00***	0.78	***1.00***	0.09	0.69	***0.98***	0.25	0.65		
LIA_NEW	-0.27	0.43	-0.09	0.81	0.48	-0.38	**0.88**	0.55	-0.18	
HS_SP	***0.94***	0.51	**0.88**	-0.17	0.40	***0.97***	-0.12	0.36	***0.91***	-0.54
HS_OLD	0.19	0.74	0.36	**0.86**	0.73	0.06	***0.93***	**0.84**	0.28	**0.88**
HS_NEW	0.45	-0.22	0.28	-0.78	-0.26	0.56	-0.75	-0.42	0.37	***-0.95***
STUMPS	0.67	0.20	0.55	-0.32	0.22	0.74	-0.41	-0.13	0.61	-0.70
DEAD	0.58	-0.07	0.43	-0.75	-0.16	0.67	-0.61	-0.19	0.50	***-0.91***
GAP	0.34	0.07	0.30	-0.36	0.16	0.36	-0.17	-0.17	0.32	-0.19
LG	-0.40	-0.39	-0.49	-0.19	-0.22	-0.34	-0.37	-0.65	-0.45	-0.29
SG	-0.10	0.34	0.01	0.59	0.52	-0.17	0.51	0.14	-0.04	0.64
SKID	0.47	0.36	0.51	0.39	0.38	0.44	0.08	0.20	0.50	0.06
GAP_SKID	0.61	0.57	0.62	-0.08	0.62	0.59	0.32	0.36	0.61	0.07
INTERIOR	0.15	0.18	0.24	0.27	0.02	0.08	0.28	0.48	0.19	0.33

with a dbh > 10 cm per ha (table 3). Compared to several other studies on species richness of 'terra firme' forest (*i.e.* forest not permanently or seasonally flooded) (appendix 2), the Greenheart forest in Guyana can be considered a mediate rich to poor tropical rain forest type. This result was also found concerning epiphytic richness of this forest type (chapter 3).

Beta diversity of different sites and plots

Spatial changes in species composition between sites and between plots ('beta diversity') is high in Greenheart dominated forest. Highest similarity values are in general between 0.5 and 0.6, and only two plots, the exploited plots of the Reserve site, show a higher similarity value (0.63). These values indicate that even plots in close vicinity share approximately just over half of the total number of species found in both plots.

Distance explained a larger part of the variance in similarity of total species

Table 4.9 (continued)
Values indicated **bold** are significant for p< 0.05, ***bold italic*** are significant for p< 0.01, ***bold italic*** (underlined) are significant for p< 0.001.

	Herbs & shrubs			Damage						
	sp	N of old	new	stumps	dead	gap area	large gap	small gap	skid trail	gap & skid
SP										
SPOLD										
SPNEW										
TRE_SP										
TRE_OLD										
TRE_NEW										
LIA_SP										
LIA_OLD										
LIA_NEW										
HS_SP										
HS_OLD	-0.09									
HS_NEW	0.65	-0.78								
STUMPS	0.80	-0.44	0.83							
DEAD	0.76	-0.67	***0.95***	0.76						
GAP	0.22	-0.14	0.43	0.43	0.28					
LG	-0.25	-0.56	0.27	0.34	0.07	0.03				
SG	-0.32	0.53	-0.47	-0.08	-0.64	0.42	0.18			
SKID	0.40	0.34	0.03	0.37	-0.07	0.38	-0.29	0.46		
GAP_SKID	0.39	0.23	0.23	0.38	0.23	0.80	-0.14	0.42	0.23	
INTERIOR	0.04	0.53	-0.40	-0.50	-0.24	-0.20	***-0.94***	-0.18	0.31	-0.16

composition between plots (table 6) and within plots (table 7) than other measured environmental variables, *e.g.* slope, logging damage, or presence of natural gaps. Plots from one site were more similar in diversity to each other (table 6), and highest similarity values within a site were always found between plots in close vicinity. The similarity of plots within one site may be partly due to site specific species. However, when we compared damage zones per plot characterized by their liana species composition, all damage zones within one site, be it gaps or interior, clustered together, and not with the same damage zones from other sites (data not shown), even when only common species are considered. Thus, local abundance of species also plays a considerable role in determining the similarity between sites and plots. On all investigated spatial scales distance dependent diversity patterns were the strongest patterns found.

This pattern of distance dependence is also reflected in the non-homogeneous behaviour of the cumulative species richness curves for all 18 unexploited one-ha

Table 4.10
Multiple linear regression for species richness after logging, new established species after logging and logging damage, for six exploited plots of Greenheart forest near Mabura Hill, Guyana.

	R-square	Intercept	SP-OLD	DEAD	Stumps	Gap	Large Gap	SKID	Gap & skid
All									
SP	0.68 *	49.75	0.77 *						
SPnew	0.68 *	103.55				-0.02 *	0.01 *		
Trees									
SP	ns								
SPnew	0.99 **	-70.17		-0.71 *		-0.03 *			0.05 *
DBH>20 cm	0.97 **	-63.24	1.37 **	-0.86 *					
5<DBH>20 cm	ns								
1<DBH<5 cm	0.99 ***	59.36					0.01 ***	-0.02 **	
DBH>1.50 m	ns								
Lianas									
SP	0.92 *	32.31	0.66 *		-0.57 *			-0.03 **	
SPnew	0.70 *	-12.65		-0.82 **		-0.01			0.02 *
DBH>10 cm	0.86 **	-8.99	0.28 **						
height>2 m	0.97 *	3.95	1.10 *		-1.49 *		0.02 *		
50 cm<height<2m	ns								
height<50 cm	0.99 **	10.17	0.27 **	-0.11 *		-0.01 *			
Undergrowth									
SP	0.93 *	16.61			0.42 **		-0.01 *		
SPnew	0.90 **	2.89		0.39 **					
height > 2 m	0.97 **	35.57		-0.26 **				-0.01 *	
50 cm<height<2m	0.97 **	35.57		-0.26 **				-0.01 *	
height<50cm	ns								

plots (Fig. 6). By adding more sites, more geographical variation is introduced, leading to higher species numbers. This is not true, however, for the undergrowth species in Greenheart forest. In the exploited sites in closer vicinity of Mabura Hill only three additional undergrowth species were found, of which two in exploited plots. Greenheart forest seems to be very uniform in undergrowth species composition, with a strong dominance of a few species (table 1, Fig. 5).

Fanshawe (1952) originally described two different associations within what we called 'Greenheart forest' in this study; namely 1) an *Eschweilera - Licania* association and 2) an *Eschweilera - Dicymbe* association. The difference between these two associations is mainly determined by the presence of a co-dominant *Dicymbe altsonii.* Fanshawe (1952) only considered a 'Chlorocardium rodiei' fasciation within the first association. In this study, the forest in Pibiri would belong to the first association,

Table 4.12

Difference between tree and liana species composition of large (≥ 300 m2) and small (< 300 m2) gaps. Difference calculated with Wilcoxon's matched pair test, and number of individuals per m2 .*: p<0.05, **: p< 0.01, ***: p<0.001.

	Waraputa			Reserve			2 km		
	T	Z	p-level	T	Z	p-level	T	Z	p-level
Large vs small									
UL Large Gap & UL Small Gap	249.5	1.07	ns	ne			393	2.86	**
Logged1 Large Gap & Logged1 Small Gap	188.5	5.69	***	286	4.18	***	121	7.25	***
Logged2 Large Gap & Logged2 Small Gap	1086	0.33	ns	9	7.58	***	924	1.86	ns
Same size, same status									
Logged1 Large Gap & Logged2 Large Gap	1261	0.68	ns	852	4.49	***	2075	0.76	ns
Logged1 Small Gap & Logged2 Small Gap	258	4.94	***	241	3.27	**	358	4.10	***
Same size, UL vs LOG									
UL Small Gap & Logged1 Small Gap	436.5	0.44	ns	ne			626	1.59	ns
UL Small Gap & Logged2 Small Gap	410	4.55	***	ne			340	4.23	***
" & " (-overlap skid)	416.5	4.05	***				586	1.91	ns
UL Large Gap & Logged1 Large Gap	293	5.28	***	622	3.09	**	467	5.88	***
" & " (-overlap skid)	341.5	4.22	***	723	2.75	*	713.5	5.18	***
UL Large Gap & Logged2 Large Gap	331.5	4.94	***	523	6.37	***	543	4.66	***
" & " (-overlap skid)	317.5	4.73	***	716.5	5.45	***	918.5	1.01	ns

association. In this study, the forest in Pibiri would belong to the first association, while the forest in the other study sites (Waraputa, Reserve and 2km) would belong to the second association. Although the presence of *Dicymbe altsonii* clearly gives a visible distinction in the field, our data does not support a division in two different types of forest. For the trees, the plots of the 2km site are even more similar in species composition to the Pibiri plots than to the other sites that also carried *Dicymbe altsonii* (Fig. 7). This coincides with the findings of Ogden (in Richards 1996), who studied forest plots at the Twenty-four Mile Reserve (Guyana) and found that within mixed forest different forest patches, either dominated by a specific tree species or without a clear dominant tree species, formed a continuum, with no clear distinctions. The differences we found between plots can be explained by geographical distance. The similarity of 2km and Pibiri (Fig. 7) is probably due to the presence of Morabukea (*Mora gonggrijpii).* at both sites, and in Pibiri this species is found in plots 11-15. The plots in Pibiri were selected based on commercially viable numbers of tree species. This is recognizable in the patterns of similarities between the Pibiri plots (Fig. 7). Plot 13 is an exceptional plot in Pibiri due to an strong dominance of Morabukea Species richness of tree species is relatively low in this plot (table 3) and species had much lower densities in this plot than in most other plots of the Pibiri site.

Table 4.11
Correlations (Pearson product moment) between logging damage and number of species and individuals for trees and lianas in 150 subplots of exploited Greenheart forest near Mabura Hill, Guyana.

	Logged subplots (N=150)							
	GAP		large gap		small gap		SKID TRAIL	
	T	L	T	L	T	L	T	L
N of species								
Total	0.07	0.27 **	0.06	0.17 *	-0.04	0.14	-0.03	0.16 *
20*	-0.14	-0.19 *	-0.13	-0.15	-0.01	-0.03	-0.12	-0.10
10*	0.03	0.29 **	0.00	0.17 *	-0.06	0.15	0.07	0.16 *
5*	0.08	0.12	0.06	-0.03	-0.01	0.18 *	-0.08	-0.13
2*	0.05	0.04	-0.01	0.00	0.15	0.10	-0.08	-0.12
N of individuals								
20*	-0.20	-0.18	-0.18	-0.14	0.14	-0.02	-0.12	-0.12
10*	0.06	0.28 **	0.08	0.14	-0.06	0.13	0.03	0.12
5*	0.21	0.15	0.17	-0.01	0.00	0.27 **	-0.06	-0.15
2*	0.09	0.04	-0.02	-0.06	0.15	0.18	-0.09	-0.12

Notes: **Large gap:** > 300 m2 **Small gap:** < 300 m2

	T	L
20*	DBH>20 cm	DBH>10 cm
10*	5 < DBH < 20 cm	DBH> 2 m
5*	2 < DBH < 5 cm	50 cm < height < 2m
2*	DBH< 2 cm height > 1.50 m	height < 50 cm

Logging induced differences

Species richness

High species richness can be found both in unexploited and exploited plots. Mean number of species per plot, however, was higher in exploited plots compared to unexploited plots, for all growth forms (table 3). Species numbers of lianas and trees were significantly positive correlated with each other (table 5), both for unexploited and exploited plots. This leads to the conclusion that based on measurements of either two, estimations can be made concerning the total species richness. Species diversity of undergrowth species, however, is only slightly correlated with total species diversity, and only for unexploited plots. Although herbs and shrubs were much less rich in species than trees and lianas, they have to be explicitly included in a sampling procedure to obtain reliable estimates on the species richness of this growth form. This growth form can be sampled on finer spatial scales than trees and lianas, because of the size of the individuals and their more even distribution in Greenheart forest (Fig. 6). We conclude that in the traditionally exploited sites especially lianas and

Table 4.11 (continued)
Values indicated **bold** are significant, *: p<0.05, **: p<0.01. (T = trees, L = lianas).

	Logged subplots (N=150)					
	GAP&SKID		OUTERGAP		INTERIOR	
	T	L	T	L	T	L
N of species						
Total	-0.13	0.06	0.03	**-0.26** **	-0.02	**-0.19** *
20*	**-0.18** *	-0.15	**0.18** *	**0.23** *	**0.18** *	0.14
10*	-0.02	0.09	-0.04	**-0.29** **	-0.03	**-0.20** *
5*	-0.11	-0.09	0.08	-0.07	-0.08	0.11
2*	0.01	0.04	0.04	-0.10	-0.08	**0.17** *
N of individuals						
20*	**-0.21**	-0.16	**0.29** **	**0.18**	0.13	**0.23**
10*	0.05	**0.24**	-0.06	**-0.28** **	-0.05	**-0.28**
5*	-0.02	-0.08	-0.08	-0.04	-0.13	0.04
2*	0.06	0.05	0.01	-0.08	-0.11	0.13

Notes: GAP&SKID: Overlap of skid trail and gap; OUTERGAP: zone from gap edge till 15 m out; INTERIOR: >15 m from gap edge

	T	L
20*	DBH>20 cm	DBH>10 cm
10*	5 < DBH < 20 cm	DBH> 2 m
5*	2 < DBH < 5 cm	50 cm < height < 2m
2*	DBH< 2 cm height > 1.50 m	height < 50 cm

trees show reactions in species composition related to logging and logging damage. Of those two growth forms, lianas showed a higher increase in species numbers and showed higher correlations with logging damage (table 3, table 11). Therefore, lianas would be most suited to use as an indicator growth form of logging damage. Specific reactions of this growth form to logging are treated in chapter 5.

Number of rare species differed between the Pibiri plots and all plots from the exploited sites, exploited and unexploited (table 4). This difference can also be attributed to geographical distance, and to species area relationships. A larger total area is expected to contain more species and the total area encompassed by the exploited sites is much larger than total area of the Pibiri plots. Logging does not seem to influence the number of rare species. These results were calculated for all growth forms together, the lianas separately (chapter 5), however, did show an decreased number of rare species after logging.

Species composition

Six years after logging significant differences in species composition between exploited and unexploited plots from one site were found (table 8). Ten years after

Table 4.13

Correlations (Pearson product moment) between logging damage and individual tree, liana, undergrowth species, for 150 exploited 20 x 20 meter subplots in exploited Greenheart forest, near Mabura Hill, Guyana. Values indicated **bold** are significant for p<0.05,***bold*** are significant for p<0.01, ***bold*** are significant for p< 0.001.

	Gap	Large gap	Small gap	Skid trail	Gap & skid	Outer gap	Interior
Trees							
Chlorocardium rodiei	-0.04	-0.05	0.00	-0.06	-0.12	0.04	0.08
Annona symphyocarpa	0.13	0.05	**0.13**	0.03	0.06	-0.01	**-0.15**
Aspidosperma exselsum	0.05	-0.05	**0.13**	-0.04	-0.11	0.05	-0.03
Cecropia sciadophylla	**0.21**	***0.30***	-0.06	-0.10	-0.03	-0.08	-0.07
Chaetocarpus schomburgkianus	-0.04	-0.01	-0.06	-0.13	0.03	**0.14**	-0.05
Clathrotropis brachypetala var. *brachypetala*	-0.05	0.02	-0.10	-0.12	0.00	-0.04	**0.14**
Dicymbe altsonii	-0.13	-0.06	-0.12	-0.07	-0.12	-0.06	***0.26***
Duguetia neglecta	-0.14	-0.13	-0.06	-0.01	-0.12	-0.05	**0.24**
Emmotum fagifolium	0.01	0.07	-0.07	***0.19***	**0.13**	-0.12	-0.03
Endlicheria punctulata	***0.20***	***0.19***	0.07	-0.06	0.06	-0.17	-0.04
Eschweilera subglandulosa	-0.06	-0.04	-0.04	-0.10	-0.17	0.08	0.10
Eugenia coffeifolia	-0.10	**-0.13**	0.01	-0.09	-0.04	***0.20***	-0.04
Goupia glabra	0.08	0.07	0.04	0.09	***0.17***	**-0.13**	-0.06
Lecythis confertiflora	**-0.14**	-0.09	-0.10	0.06	-0.05	**-0.14**	***0.27***
Maburea trinervis	-0.06	-0.03	-0.06	-0.08	-0.11	-0.04	***0.18***
Oxandra asbeckii	-0.10	0.00	**-0.16**	-0.11	-0.11	-0.12	***0.31***
Quiina guianensis	0.01	-0.03	0.07	-0.09	-0.02	**0.15**	-0.11
Tapura guianensis	0.02	-0.01	0.04	-0.04	**-0.14**	-0.01	0.07
Unonopsis glaucopetala	-0.11	-0.07	-0.07	0.00	**-0.14**	0.05	0.11
Lianas							
Anemopaegma parkeri	**0.14**	0.10	0.10	-0.06	***0.18***	-0.06	**-0.13**
Aristolochia daemoninioxia	**0.13**	0.10	0.04	-0.06	-0.01	0.01	-0.10
Coccoloba parimensis	**0.14**	**0.12**	-0.01	0.00	0.00	-0.08	-0.06
Curarea candicans	0.05	-0.01	**0.12**	0.06	**0.14**	-0.06	-0.08
Machaerium madeirense	0.06	0.03	0.06	**0.12**	0.07	0.01	-0.09
Machaerium myrianthum	**0.13**	***0.17***	-0.07	-0.01	***0.18***	**-0.14**	-0.06
Maripa scandens	***0.18***	**0.13**	0.07	0.06	0.08	-0.02	***-0.21***
Mascagnia sepium	0.06	-0.06	-0.01	-0.05	**0.14**	-0.07	-0.02
Memora racemosa	0.11	***0.20***	-0.02	-0.05	-0.01	-0.05	-0.03
Moutabea guianensis	-0.04	-0.02	-0.01	**0.17**	-0.04	-0.05	0.03
Passiflora spp.	0.09	***0.18***	-0.08	***0.26***	***0.27***	**-0.12**	***-0.19***
Paullinia pachycarpa	0.09	0.11	-0.05	-0.04	-0.03	0.01	-0.06
Pinzona coriacea	***0.19***	***0.21***	0.03	**0.19**	***0.37***	***-0.19***	***-0.24***
Rourea pubescens	0.10	**0.13**	0.01	0.06	0.07	-0.03	**-0.13**
Strychnos bredemeyeri	**0.13**	0.10	0.02	-0.03	**0.12**	-0.10	-0.07
Tontelea coriacea	0.03	0.09	-0.04	**0.15**	-0.02	-0.01	-0.08

Table 4.13 (continued)

	Gap	Large gap	Small gap	Skid trail	Gap & skid	Outer gap	Interior
Herbs & shrubs							
Bactris balanophora	***0.20***	0.14	0.11	-0.06	-0.06	-0.06	-0.09
Ischnosiphon gracilis	0.08	0.13	-0.04	0.15	***<u>0.28</u>***	***-0.22***	-0.16
Miconia gratissima	0.13	***0.22***	-0.07	***<u>0.39</u>***	-0.06	***<u>-0.24</u>***	-0.09
Miconia punctata	***<u>0.27</u>***	***<u>0.28</u>***	0.06	0.06	0.04	***<u>-0.25</u>***	-0.10
Miconia sp1.	-0.11	-0.05	-0.09	***0.21***	***0.22***	-0.08	-0.10
Renealmia orinocensis	-0.04	0.03	-0.09	0.11	***<u>0.27</u>***	-0.10	-0.13
Scleria sp.	-0.01	***0.20***	0.02	-0.06	0.03	0.05	-0.05

Table 4.14
Amount of large gap and small gap area for nine plots in Greenheart dominated forest, near Mabura Hill, Guyana. All plots were logged with conventional logging methods. Gap area is without an uncertainty zone of 0.5 meter.

		Large gap (m2)	Small gap (m2)	Total
Waraputa	**Unlogged**	902.5	576.0	1478.5
	LOG1	717.0	2637.8	3354.8
	LOG2	1508.5	1706.1	3214.6
Reserve	**Unlogged**	973.8	0.0	973.8
	LOG1	912.3	1649.0	2561.3
	LOG2	936.2	2674.5	3610.7
2 km	**Unlogged**	811.5	770.0	1581.5
	LOG1	1300.4	2487.0	3787.4
	LOG2	923.4	2232.3	3155.7

there existed a significant difference in species composition between large gaps in the unexploited plot and both exploited plots (table 12), and between small gaps in the unexploited plot and in one of the exploited plots. Thus, both logging and gaps size influence species differences between plots. The differences in species composition between gaps of the same size in exploited and unexploited plots at the 2km site could possibly be explained by the geographical distance between the plots. This is contradicted by the fact that on the level of the whole one-ha the difference is no longer significant (table 8). The difference may also be explained by the presence of overlapping gap and skid trail area in the exploited plots. Indeed, when this overlapping area is excluded from the analyses for the 2km site, the distinction between the species composition in large gaps in the unexploited plot and one of the exploited plots, and in small gaps, is no longer valid. However, this effect of exclusion was not found at the other sites. So, overlapping gap and skid trail can only be part of the explanation. Also the total amount of large and small gap area that occurred in the different plots (table 14) may be important. Although species composition is evaluated per square meter, due to species-area relationships, a larger total area is expected to contain more species. The large gaps in the Reserve site, however, were all of comparable sizes, and did differ significantly in all combinations. In conclusion, logging introduced significantly different species compositions between gaps of similar sizes occurring in exploited and unexploited plots of the same site. This distinction can only in part be explained by overlapping gap and skid trail area that is included in the gaps in exploited plots.

Growth forms

1 Trees

Total number of tree species after logging did not show a significant relation with combined species richness and logging damage variables (table 10), while it showed a higher (but not significant) correlation with the number of newly established species after logging than with the number of tree species already present before the logging (table 9). These results seem to indicate that the majority of tree species before logging is still present after logging, and that shifts in species richness are mainly due to newly established species. We assumed the size class of trees with a dbh between 1 and 5 cm to represent the cohort of individuals that established after logging, an assumption supported by table 10. This size class showed positive correlation with total amount of gap area, large gap area, and outer gap zone (table 11), while it was negatively correlated with combined gap and skid trail area. This seems to be in contrast with the positive correlation between number of newly established species and this same overlapping area, obtained in the multiple linear regression. However, individuals of newly established species are only a subset of individuals occurring in this size class. For all species, including those that were present before logging, new establishment sites are created, both inside gaps and in the areas outside gaps with higher light intensities. The sites with high disturbance, the overlapping gap and skid trail area, are mainly new establishment sites for the invading species after logging.

Differences in species composition between plots (Fig. 7) and subplots (table 7) was only partly dependent on distance between the (sub)plots, and on the level of one-ha plots no direct relation with logging damage could be distinguished. Other factors, including chance, seem to regulate differences in species composition of trees.

It should be noted that actual species richness of trees could be higher than the values listed. Identifications of trees were mostly made with the assistance of tree spotters, and local names were translated into scientific names. The identifications in the field were checked with voucher specimens and some local names proved to encompass different species. When possible, these species were recognized separately, otherwise species were lumped (*e.g.*, the vernacular name Aruadan encompasses at least two different species of *Sloanea*).

2 Lianas

Total species richness of lianas on the level of one-ha plots showed a significant positive correlation with the number of species present before the logging, and a negative correlation with logging damage, expressed as number of harvested trees and skid trail area. This seems to indicate that liana species disappear together with the harvested trees. This could be the large lianas growing within the harvested trees and their immediate surroundings. Within the unexploited plots of Pibiri we found that a considerable part (60%) of the large liana species occur in low densities (table 2), this enhances the chance that a large liana individual belongs to a rare species (rare on the level of one-ha) and thus disappears after logging. Newly established species are mainly positively correlated with overlapping gap and skid trail area, while negatively correlated with total gap area and number of dead standing trees. The amount of skid trail area is also an important variable explaining variance in species numbers between the plots.

3 Herbs, shrubs, and small palms

The undergrowth species showed a much finer-grained mosaic of occurrences compared with the trees and lianas. There was a large group of species (17 spp., 43.6%) growing in more than 10 plots, and even 7.7% of the species occur in all plots. For lianas and trees the comparable values were 27.9% (2.2% in all) and 28.7% (2.4% in all), respectively. Of the 27 species found in the exploited plots, only one was not found in the undisturbed plots (appendix 1). These result indicate that undergrowth species are more evenly distributed over the forest floor and they seem to be relatively shade tolerant. Already with 11 plots, 90% of all undergrowth species were sampled (Fig. 6). For unexploited Greenheart forest we listed 12 undergrowth species that contributed to more than 90% of all individuals (table 1). Of those 12 species, only two showed a significant correlation with logging damage classes (table 13) or increased light classes (gap, large gap, small gap). Topographic variation between and within the plots was slight or absent, and all plots were situated on brown sand soils. Possible differences may be expected in soil moisture availability which may effect understorey diversity (Dirzo *et al.* 1992; Newbery *et al.* 1996; Poulsen 1996).

The different growth forms all showed their own responses to logging, at different spatial scales. Undergrowth species showed a rather even distribution throughout all investigated sites and plots and showed no difference in species composition in exploited or unexploited forest. Liana species had a much more patchy distribution and differences in species composition could be largely explained by geographical distance, and secondly by logging damage. Tree species also showed a patchy distribution, but on a larger spatial scale than the lianas. Thus, the species composition of undergrowth and lianas (and epiphytes, chapter 3) seem to be more determined by their environment than the species composition of trees.

Lianas are the most responsive growth form to logging-induced changes in the environment. The types of responses are more detailed discussed in chapter 5. Composition of tree species after logging seems mainly determined by newly established species after logging. The more overlapping gap and skid trail area created during logging, the more new, invading, species arrive (both trees and lianas).

As regards to forest management, this would indicate that in order to maintain the characteristic species composition of Greenheart forest, the extent of overlapping gap and skid trail area should be kept at a minimum. It should be noted, however, that these results were obtained from a comparison of an unexploited referential plot with a exploited plot and not from censuses in the same plot before and after logging. The evaluation of reduced-impact logging, as scheduled for the future by the Tropenbos Foundation, is needed for further evaluations of the responses of the plant community to logging.

Acknowledgements

The authors would like to thank Demerara Timbers Ltd. for their hospitality and cooperation during our subsequent stays in the field. Special thanks are due to three excellent Guyanese tree spotters: I. Johnson, A. Braithwaite, and S. Roberts, and all Guyanese field assistents. Furthermore, C. Brils, D. Bus, M. de Lind Van Wijngaarden, K. Eichhorn, M. Elsinga, B. Gravendeel, J. Van Dam, C. Van der Kooij, E. Laan, B. Laumans, N. Raes, B. Robers, F. Slik, A. Stroo, E. Vanmechelen, and L. Versteeg for their help in fieldwork. P. Hogeweg and M.J.A. Werger are acknowledged for their comments on the first version of this manuscript. This study was carried out in the framework of the Tropenbos Guyana Programme and funded by the Tropenbos Foundation, Wageningen, the Netherlands, the Utrecht University, and the EU.

Appendix 4.1

List of species occurring in 24 one-ha plots of Greenheart forest, near Mabura Hill, Guyana. Pib= Pibiri, War= Waraputa, Res= Reserve, 2km= '2km'. Local names in capitals are locally used, those with small print were added for further distinction or using Mennega *et al.* (1988).

Species	Local name	Growth form	dbh > 10 cm	Pib	War	Res	2km
Anacardiaceae							
Anacardium giganteum Hancock ex Engl.	UBUDI	T		0	0	1	0
Tapirira obtusa (Benth.) J.D. Mitchell	DUKA	T	*	1	1	1	1
Annonaceae							
Anaxagorea dolichocarpa Sprague & Sandw.	KURIHIKOYOKO	T		1	1	1	1
Annona symphyocarpa Sandw.	DURU	T		0	1	1	1
Bocageopsis multiflora (Mart.) R.E. Fr.	ARARA, Fine Leaf	T	*	1	0	0	1
Duguetia neglecta Sandw.	YARI YARI	T		1	1	1	1
Duguetia yeshidan Sandw.	YESHIDAN	T		1	0	0	1
Guatteria sp1.	ARARA, Smooth Skin	T	*	0	1	1	1
Guatteria sp2.	YARI YARI, Black	T		1	1	1	1
Oxandra asbeckii (Pulle) R.E. Fr.	KARISHIRI	T	*	1	1	1	1
Unonopsis glaucopetala R.E. Fr.	ARARA, Broad leaf	T	*	0	1	1	1
Unonopsis sp1.	ARARA	T		1	1	0	0
Xylopia nitida Dunal	KUYAMA, White	T		1	1	0	1
Xylopia pulcherrima Sandw.	KUYAMA, Red	T		1	1	1	0
sp1.		T		0	1	0	0
Apocynaceae							
Ambelania acida Aubl.	MAKORIRO	T		1	0	0	0
Aspidosperma album (Vahl) Benoist ex Pichon cf.	SHIBADAN, Smooth leaf	T	*	1	0	0	0
Aspidosperma cruentum Woodson cf.	SHIBADAN, Boat leaf	T	*	1	0	1	0
Aspidosperma exselsum Benth.	YARURU	T	*	1	1	1	1
Aspidosperma sp.	SHIBADAN	T		0	1	1	1
Forsteronia acouci (Aubl.) A. DC.		L (W)		1	0	1	1
Forsteronia gracilis (Benth.) Muell. Arg.	Makwariballi	L (W)		1	1	0	1
Forsteronia guyanensis Muell. Arg.	Makwariballi	L (W)		1	0	1	1
Forsteronia schomburgkii A. DC. cf.		L (W)		1	0	1	1
Geissospermum sericeum (Sagot) Benth. & Hook. f.	MANYOKINABALLI	T	*	1	1	0	0
Himathanthus articulatus (Vahl) Woodson	MABWA	T	*	1	0	0	0
Macoubea guianensis Aubl.	ROKOROKO	T		0	1	1	0
Odontadenia amazonica (Stadelm.) Malme cf.		L (W)		1	0	0	0
Odontadenia puncticulosa (A. Rich.) Pulle		L (W)		1	1	1	0
Odontodenia geminata (Roem. & Schult.) Muell. Arg. cf.		L (W)		1	0	0	1

Species	Local name	Growth form	dbh > 10 cm	Pib	War	Res	2km
Parahancornia fasciculata (Poir.) Benoist ex Pichon	DUKALI	T	*	1	0	1	0
Tabernaemontana cerea (Woodson) Leeuwenberg	PERO-ISHI-LOKODO, Rough bark	T		0	1	1	1
Tabernaemontana disticha DC.	PERO-ISHI-LOKODO, Smooth bark	T		0	1	1	1
Tabernaemontana undulata Vahl	PERO-ISHI-LOKODO, Marginal vein	T		1	1	1	1
Aquifoliaceae							
Ilex martiniana D. Don	KAKO	T		1	0	0	0
Araceae							
Heteropsis flexuosa (Kunth) Bunting	NIBI	HE		1	1	1	1
Philodendron rudgeanum Schott	ICANA	HE		1	1	1	1
Rhodospatha venosa Gleason		HE		1	0	0	0
Aristolochiaceae							
Aristolochia consimilis Mast.		L (V)		1	0	0	0
Aristolochia daemoninoxi Mast.	BOYARI	L (W)		1	0	1	1
Aristolochia rugosa Lam.		L (V)		0	1	0	0
Bignoniaceae							
Anemopaegma oligoneuron (Sprague & Sandw.) A. Gentry		L (W)	*	1	1	1	1
Anemopaegma parkeri Sprague		L (W)		1	1	1	1
Arrabidaea egensis Bureau ex K.Schum.	Hakiaballi	L (W)		0	0	0	1
Arrabidaea fanshawei Sandw.	Hakiaballi	L (W)		1	0	0	0
Arrabidaea mollis (Vahl) Bureau ex K.Schum.	Hakiaballi	L (W)	*	1	0	1	1
Clytostoma binatum (Thunb.) Sandw.		L (W)		1	1	0	1
Clytostoma sciuripabulum Bureau & K. Schum.		L (W)		1	0	1	1
Cydista aequinoctialis (L.) Miers		L (W)		1	0	0	0
Distictella elongata (Vahl) Urb.	Kamoro	L (W)		1	0	0	0
Jacaranda copaia (Aubl.) D. Don subsp. copaia	FUTUI	T	*	1	0	1	1
Memora bracteosa (DC.) Bureau & K. Schum. cf.		L (W)		1	0	0	0
Memora flavida (DC.) Bureau & K. Schum.		L (W)	*	0	1	1	1
Memora moringifolia (Miq.) Pulle		L (W)		1	1	1	1
Memora racemosa A. Gentry		L (W)		1	0	1	1
Memora schomburgkii (DC.) Miers		L (W)		0	0	0	1
Mussatia prieuriei (DC.) Bureau ex K. Schum.	Kamoro	L (W)		1	0	0	1
Pleonotoma albiflora (Salzm. ex DC.) A. Gentry		L (W)		1	1	1	1
Roentgenia sordida (Bureau & K. Schum.) Sprague & Sandw.		L (W)		1	0	0	0
Schlegelia violacea (Aubl.) Griseb.	Bultatakubia	HE		1	1	1	0

Species	Local name	Growth form	dbh > 10 cm	Pib	War	Res	2km
Bombacaceae							
Catostemma commune Sandw.	BAROMALLI, Swamp	T	*	1	0	1	0
Catostemma fragrans Benth.	BAROMALLI, Sand	T	*	1	1	1	1
Boraginaceae							
Cordia exaltata Lam. var. exaltata	TABLE TREE	T	*	1	0	1	1
Cordia nodosa Lam.	HURUEREROKO	T		1	1	1	1
Burmanniaceae							
Gymnosiphon breviflorus Gleason		U (H)		1	1	0	1
Hexapterella gentianoides Urb.		U (H)		1	0	0	0
Burseraceae							
Protium decandrum (Aubl.) Marchand	KUROKAI	T	*	1	1	1	1
Protium guianense (Aubl.) Marchand var. guianense	HAIAWA	T	*	1	1	1	1
Protium tenuifolium (Engl.) Engl.	HAIAWABALLI	T	*	1	0	0	1
Trattinickia demerarae Sandw.	ULU rsk	T		1	0	0	0
Trattinickia rhoifolia Willd.	ULU ssk	T		1	0	0	0
Trattinickia sp.	ULU	T	*	0	1	1	1
Cecropiaceae							
Cecropia angulata I.W. Bailey cf.	CONGO PUMP, White	T		0	0	1	1
Cecropia obtusa Trécul.	CONGO PUMP, Common	T		1	0	0	0
Cecropia sciadophylla Mart.	CONGO PUMP, Long leaf	T		0	0	0	1
Coussapoa microcephala Trécul	Mabakubia	HE	*	1	1	1	1
Celastraceae							
Goupia glabra Aubl.	KABUKALLI	T	*	1	1	1	1
Maytenus myrsinoides Reissek	KAIARIMA	T		1	0	0	0
Chrysobalanaceae							
Couepia exflexa Fanshawe & Maguire cf.	BOKOTOKON	T		1	0	1	0
Hirtella caduca Fanshawe & Maguire cf.	FIRE TREE	T		1	0	0	0
Hirtella racemosa Lam. var. hexandra (Willd. ex. Roem. & Schult.) Prance	BOKOTOKON	T	*	1	1	1	1
Licania alba (Bernouilli) Cuatrec.	KAUTABALLI	T	*	1	1	1	1
Licania canescens Benoist cf.	MARISHIBALLI	T	*	1	0	1	1
Licania heteromorpha Benth. heteromorpha	BURUBURULI	T	*	1	0	0	1
Licania heteromorpha Benth. var. perplexans Sandw.	KAIRIBALLI	T	*	1	1	1	1
Licania hypoleuca Benth.	UNIKIAKIA	T	*	1	0	1	1
Licania laxiflora Fritsch or L. persaudii Fanshawe & Maguire	KAUTA	T	*	1	1	1	1
Licania micrantha Miq. cf.	MARISHIBALLI, Red	T	*	1	0	0	1

Species	Local name	Growth form	dbh > 10 cm	Pib	War	Res	2km
Licania sp1	KONOKO	T	*	1	0	0	0
Licania sp2	Licania sp.	T		0	0	1	0
Licania sp3	MARISHIBALLI, Broad leaf	T		0	0	0	1
Parinari campestris Aubl.	BURADA	T	*	1	0	0	1
sp.		T	*	0	0	1	0
Combretaceae							
Terminalia amazonia (J.F. Gmel.) Exell	FUKADI	T	*	1	1	1	1
Compositae							
Mikania gleasonii B.L. Rob.		L (W)		1	0	1	1
Mikania sp.		L (V)		0	0	1	0
Connaraceae							
Cnestidium guianense (Schellenb.) Schellenb.		L (W)		1	0	0	0
Connarus coriaceus Schellenb.		L (W)/T		1	0	1	1
Connarus erianthus Benth. ex Baker var. stipitatus Forero		L (W)		1	1	1	1
Connarus megacarpus S.F. Blake	Supple Jack, LIGHT BROWN BELLY	L (W)		1	1	1	1
Connarus perrottetii (DC.) Planch. var. rufus Forero	BROWN BELLY	L (W)	*	1	1	1	1
Connarus punctatus Planch		L (W)		0	1	0	1
Pseudoconnarus macrophyllus (Poepp. & Endl.) Radlk.		L (W)		1	0	0	0
Rourea induta Planch. var. induta	Supple Jack	L (W)		0	1	1	1
Rourea ligulata Baker cf..	Supple Jack	L (W)		1	1	1	1
Rourea pubescens (DC.) Radlk. var. spadicea (Radlk.) Forero	Supple Jack	L (W)	*	1	1	1	1
Convolvulaceae							
Dicranostyles guianensis Mennega		L (W)		0	0	1	0
Lysiostyles scandens Benth.		L (W)		1	1	1	0
Maripa glabra Choisy cf.		L (W)		1	0	1	0
Maripa scandens Aubl.	MONKEY SYRUP	L (W)	*	1	1	1	1
Cucurbitaceae							
Cayaponia ophthalmica R.E. Schultes		L (V)		1	0	1	0
Cayaponia selysioides C. Jeffrey		L (V)		1	0	0	0
Gurania bignoniacea (Poepp. & Endl.) C. Jeffrey		L (V)		0	0	1	1
Gurania spinulosa (Poepp. & Endl.) Cogn.		L (V)		1	0	1	1
Helmontia leptantha (Schltd.) Cogn.		L (V)		1	1	1	1
Cyclanthaceae							
Evodianthus funifer (Poit.) Lindm. spp. trailanus (Drude) Harling		HE		1	0	0	1

Species	Local name	Growth form	dbh > 10 cm	Pib	War	Res	2km
Cyperaceae							
Calyptrocarya glomerulata (Brogn.) Urb. var. glomerulata		U (H)		1	1	1	1
Diplasia karataefolia L.C. Rich.		U (H)		1	1	1	1
Scleria sp.		U (H)		1	1	1	1
Dichapetalaceae							
Tapura guianensis Aubl.	WAIABALLI	T	*	1	1	1	1
Dichapetalum pedunculatum (DC.) Baill.	Kanakudiballi	L (W)		1	1	1	0
Dichapetalum rugosum (Vahl) Prance	Kanakudiballi	L (W)		1	1	1	1
Dilleniaceae							
Doliocarpus brevipedicellatus Garcke subsp. brevipedicellatus	KAPADULA, Red	L (W)		1	1	0	0
Doliocarpus guianensis (Aubl.) Gilg.	KAPADULA	L (W)		1	1	1	1
Doliocarpus macrocarpus Mart. ex Eichler	KAPADULA, Hairy	L (V)		1	1	1	0
Doliocarpus major J.F. Gmel. subsp. major	KAPADULA	L (W)		1	0	0	0
Doliocarpus paraensis Sleumer	KAPADULA	L (W)		0	1	1	1
Pinzona coriacea Mart. & Zucc.	FIRE ROPE	L (W)	*	1	1	1	1
Tetracera asperula Miq.	KAPADULA	L (W)		1	1	0	0
Tetracera volubilis L. subsp. volubilis	KAPADULA, White	L (W)	*	1	1	1	1
Dioscoreaceae							
Dioscorea dodecaneura Vell.	WILD YAM	L (V)		1	0	1	0
Dioscorea megacarpa Gleason cf.	WILD YAM	L (V)		0	1	0	0
Ebenaceae							
Diospyros dichroa Sandw.	BARABARA, Fine leaf	T		0	1	1	0
Diospyros ierensis Britton	BARABARA	T		0	1	1	0
Elaeocarpaceae							
Sloanea eichleri K. Schum. cf.	ARUADAN, Broad leaf	T		0	1	0	0
Sloanea guianensis (Aubl.) Benth.	ARUADAN	T	*	1	1	1	1
Sloanea sp.	ARUADAN, Small leaf	T		1	0	0	1
Erythroxylaceae							
Erythroxylum sp.	Cocaine tree	T		0	0	1	0
Euphorbiaceae							
Alchornea sp.	KANAKUDIBALLI	T		1	0	0	0
Aparisthmium cordatum (A. Juss.) Baill.	MABABALLI	T	*	1	0	0	0
Chaetocarpus schomburgkianus (Kuntze) Pax & K. Hoffm.	RURI	T	*	1	1	1	1
Croton sp.		L (V)		1	0	0	0
Dalechampia olympiana Kuhlm. & Rodr.		L (V)		1	0	0	0
Dalechampia parvibracteolata Lanj. cf.		L (V)		1	0	0	0
Glycydendron amazonicum Ducke	DEVILDOOR TREE	T	*	1	0	0	0

Species	Local name	Growth form	dbh > 10 cm	Pib	War	Res	2km
Hyeronima alchorneoides Allemão var. alchorneoides	SURADAN	T		1	0	0	0
Maprounea guianensis Aubl.	AWATI	T		0	0	1	1
Sandwithia guyanensis Lanj.	'UNKNOWN A'	T		1	1	1	1
Flacourtiaceae							
Carpotroche surinamensis Uitt.		T		0	1	1	1
Casearia sp. / Ryania speciosa Vahl var. tomentosa (Miq.) Monachino	KIBIHIDAN	T		1	1	1	1
Gentianaceae							
Irlbachia purpurascens (Aubl.) Maas		U (H)		1	0	0	0
Voyria corymbosa Splitg. subsp.alba (Standley) Ruyters & Maas		U (H)		1	1	1	1
Voyria rosea Aubl.		U (H)		1	0	0	0
Voyriella parviflora (Miq.) Miq.		U (H)		1	0	0	0
Gesneriaceae							
sp1		U (H)		1	1	1	1
sp2		U (H)		0	1	0	0
Gnetaceae							
Gnetum nodiflorum Brogn.	Bellbirds heart	L (W)		1	1	1	1
Gnetum urens (Aubl.) Blume	Bellbirds heart	L (V)		0	1	1	1
Guttiferae							
Clusia grandiflora Splitg.	KUFA	HE	*	1	1	1	1
Clusia myriandra (Benth.) Planch. & Triana	KUFA	HE		1	0	0	1
Clusia pana-panari (Aubl.) Choisy	KUFA	HE	*	1	0	0	1
Rheedia benthamiana Planch. & Triana	ASASHI	TL		1	0	0	1
Tovomita spp.	AWASOKULE	T	*	1	1	1	0
Vismia guianensis (Aubl.) Choisy	BLOODWOOD, Small leaf	T		0	0	1	1
Vismia macrophylla Kunth cf..	BLOODWOOD, Broad leaf	T		0	0	1	1
sp1		T		0	0	0	1
sp2		T		0	0	0	1
Hippocrateaceae							
Cheiloclinium cognatum (Miers) A.C. Sm.	MONKEY SYRUP	T/L (W)		1	0	0	0
Cheiloclinium hippocrateoides (Peyr.) A.C. Sm.		L (W)		0	1	0	0
Peritassa laevigata (Hoffmanns.) A.C. Sm. cf.		L (W)		1	0	0	0
Prionostemma aspera (Lam.) Miers		L (W)		1	0	0	0
Salacia multiflora (Lam.) DC. subsp. multiflora		L (W)		1	0	1	1
Tontelea attenuata Miers		L (W)	*	1	1	1	1
Tontelea coriacea A.C. Sm. cf.		L (W)	*	1	1	1	1
sp1		L (W)		1	0	0	0

Species	Local name	Growth form >	dbh 10 cm	Pib	War	Res	2km
sp2		L (W)		0	0	0	1
Humiriaceae							
Humiria balsemifera (Aubl.) A. St. Hil. var. floribunda (Mart.) Cuatrec.	TAURONIRO	T		0	1	1	1
Sacoglottis guianensis Benth. var. guianensis	DUKURIA	T	*	1	1	0	1
Icacinaceae							
Emmotum fagifolium Desv. ex Hamilton	MANOBODIN	T		1	0	1	1
Leretia cordata Vell.		L (W)		1	0	0	0
Lauraceae							
Aniba citrifolia (Nees) Mez	GALE, Almond	T		1	1	1	1
Aniba excelsa Kosterm.	GALE, Greenheart	T	*	1	0	1	0
Aniba hypoglauca Sandw.	SILVERBALLI, Yellow	T	*	1	0	1	1
Aniba kappleri Mez	GALE, Ginger	T		1	1	0	1
Chlorocardium rodiei (Schomb.) Rohwer, Richter & van der Werff	GREENHEART	T	*	1	1	1	1
Endlicheria punctulata (Mez) C.K. Allen	YEKORO, SWIZZLE STICK	T		1	1	1	1
Licaria cannella (Meissn.) Kosterm.	SILVERBALLI, Brown	T	*	1	1	0	0
Ocotea floribunda (Sw.) Mez cf.	SILVERBALLI, Pear Leaf	T	*	1	0	0	0
Ocotea guianensis Aubl. var. guianensis	SHIRUA	T	*	1	0	0	0
Ocotea puberula (Rich.) Nees	SILVERBALLI, Kereti	T	*	1	1	1	1
Ocotea spp.	SILVERBALLI, Others	T	*	0	0	1	0
sp1		T		0	1	0	0
Lecythidaceae							
Couratari guianensis Aubl.	WADARA	T	*	1	0	0	1
Eschweilera alata A.C. Sm.	KAKARALLI, Guava skin	T		0	1	1	1
Eschweilera coriacea (A. DC.) Mori / Eschweilera decolorans Sandw.	KAKARALLI, Fine Leaf Thick Skin	T	*	1	1	1	0
Eschweilera sagotiana Miers	KAKARALLI, Black	T	*	1	1	1	1
Eschweilera wachenheimii (Benoist) Sandw.	KAKARALLI, Smooth leaf	T	*	1	0	0	0
Lecythis confertiflora (A.C. Sm.) Mori	WIRIMIRI	T	*	1	1	1	1
Lecythis corrugata Poit. subsp. corrugata	WINA	T		1	1	0	0
Lecythis holcogyne (Sandw.) Mori	HAUDAN	T	*	1	0	1	1
Lecythis zabucajo Aubl.	MONKEY POT	T	*	1	1	1	1
Leguminosae/Caesalpinoideae							
Bauhinia guianensis Aubl. var. guianensis	MONKEY LADDER	L (W)	*	1	0	0	1
Bauhinia scala-simiae Sandw.	MONKEY LADDER	L (W)		0	0	0	1
Bauhinia surinamensis Amsh. cf.		L (W)		1	0	0	0
Bauhinia sp1		L (W)		0	0	0	1

Species	Local name	Growth form >	dbh 10 cm	Pib	War	Res	2km
Bauhinia sp2		L (W)		1	0	0	0
Cassia cowanii Irwin & Barneby var. guianensis (Sandw.) Irwin & Barneby	WARUA, MONKEY MALASSES	T	*	1	0	0	0
Chamaecrista adiantifolia (Benth.) Irwin & Barneby var. pteridophylla (Sandw.) Irwin & Barneby	IMIRIMIABALLI	T		0	1	1	1
Chamaecrista apocouita (Aubl.) Irwin & Barneby	APOKAITO	T	*	0	1	1	1
Dicymbe altsonii Sandw.	CLUMP WALLABA	T	*	0	1	1	1
Eperua falcata Aubl.	WALLABA, Soft	T	*	1	1	1	1
Eperua grandiflora (Aubl.) Benth.	ITURI WALLABA	T		0	1	0	0
Eperua rubiginosa Miq. var. rubiginosa	WATAPA	T		0	1	1	1
Hymenaea courbaril L. var. courbaril	LOCUST	T	*	1	0	1	1
Mora excelsa Benth.	MORA	T		1	0	0	0
Mora gonggrijpii (Kleinh.) Sandw.	MORABUKEA	T	*	1	1	1	1
Peltogyne venosa (Vahl) Benth. subsp. densiflora (Spruce ex Benth.) M.F. da Silva	PURPLEHEART	T	*	1	1	1	1
Peltogyne sp.	PURPLEHEART, Saka	T	*	1	0	0	0
Sclerolobium guianense Benth. var. guianense	KADITIRI	T	*	1	0	1	1
Tachigali rusbyi Harms cf.	YAWAREDAN	T	*	1	0	1	0
Vouacapoua macropetala Sandw.	SAREBEBEBALLI	T	*	1	0	1	1
Leguminosae/Mimosoideae							
Aberema jupunba (Willd.) Britton & Killip var. trapezifolia (Vahl) Barneby & Grimes	HURUASA	T	*	1	0	0	1
Balizia pedicellaris (DC.) Barneby & Grimes	MANARIBALLI, Red Skin	T	*	1	0	0	0
Enterolobium barnebianum Mesquita cf..	DEVIL GRANDFATHER's EAR	T	*	1	0	0	0
Inga alba (Sw.) Willd.	MAPOROKON	T	*	1	1	1	1
Inga lateriflora Miq.	WARAKOSA, Long leaf	T		0	0	0	1
Inga sp1	WARAKOSA, Fine leaf	T		0	0	1	1
Inga sp2	WARAKOSA, Round leaf	T		0	1	1	1
Inga sp3	WARAKOSA, Small leaf	T		0	0	0	1
Inga spp.	WARAKOSA	T	*	1	1	0	1
Parkia ulei (Harms) Kuhlm. var. surinamensis Kleinh.	UYA	T	*	1	0	0	0
Pentaclethra macroloba (Willd.) Kuntze	TRYSIL	T	*	1	0	1	1
Pseudopiptadenia suavolens (Miq.) Grimes	MANARIBALLI , Like	T	*	0	0	0	1
Zygia racemosa (Ducke) Barneby & Grimes	TURELI	T	*	1	0	1	1

Species	Local name	Growth form >	dbh 10 cm	Pib	War	Res	2km
Leguminosae/Papilinoideae							
Clathrotropis brachypetala (Tul.) Kleinh. var. brachypetala	AROMATA	T	*	1	1	1	1
Clathrotropis macrocarpa Ducke	AROMATA, Like	T		0	1	1	1
Clitoria sagotii Fantz. var. sagotii		L (W)		0	0	1	0
Clitoria sagotii Fantz var. canaliculata Fantz		L (W)		0	1	0	0
Dioclea elliptica R.H. Maxwell		L (W)		1	0	0	0
Dioclea scabra (Rich.) R.H. Maxwell		L (W)		1	0	0	0
Diplotropis purpurea (Rich.) Amsh. var. purpurea	TATABU	T	*	1	1	0	0
Dipteryx odorata (Aubl.) Willd.	TONKA BEAN	T	*	1	0	0	0
Hymenolobium flavum Kleinh. cf.	DARINA	T	*	1	0	0	0
Hymenolobium petraeum Ducke cf.	KORAROBALLI	T	*	1	0	0	0
Lonchocarpus negrensis Benth.	Haiariballi	L (W)	*	1	0	1	1
Machaerium madeirense Pittier		L (W)	*	1	0	1	1
Machaerium mutisii Killip ex Rudd		L (W)		1	0	0	1
Machaerium myrianthum Spruce ex Benth.	Hold-me-back, Wait-a-second	L (W)		1	0	1	1
Machaerium oblongifolium Vogel		L (W)		1	0	0	0
Machaerium quinata (Aubl.) Sandw. var. quinata	Bohoribada	L (W)		1	1	1	0
Machaerium sp.		L (W)		1	0	0	0
Ormosia coccinea (Aubl.) Jackson	BARAKARO	T	*	1	1	1	1
Ormosia coutinhoi Ducke	KOROKORORO	T	*	1	1	1	0
Swartzia benthamiana Miq. var. benthamiana	ITIKIBOROBALLI	T		1	0	0	0
Swartzia guianensis (Aubl.) Urb.	ITIKIBOROBALLI, Long small leaves	T		0	1	1	1
Swartzia jenmanii Sandw.	PARAKUSAN	T	*	1	1	1	1
Swartzia latifolia Benth. var. sylvestris Cowan	ITIKIBOROBALLI, Broad leaf	T		0	1	0	0
Swartzia leiocalycina Benth.	WAMARA	T	*	1	1	1	1
Swartzia oblanceolata Sandw.	SEREBEDAN	T	*	1	1	1	0
Swartzia sp. (prob. sp. nov.)	ITIKIBOROBALLI, Cauliflorous	T		0	1	1	1
Swartzia sp1	ITIKIBOROBALLI, Broad leaf	T		0	0	1	0
Swartzia sp2	ITIKIBOROBALLI, Hairy leaves	T		0	1	0	0
Swartzia sp3	ITIKIBOROBALLI, Round leaf	T		0	0	1	0
Swartzia sp4	ITIKIBOROBALLI, Small leaf	T		0	0	1	0
Swartzia spp.	ITIKIBOROBALLI, Others	T		1	1	1	1

Species	Local name	Growth form	dbh > 10 cm	Pib	War	Res	2km
Linaceae							
Hebepetalum humiriifolium (Planch.) Benth.	SHIBALLIDAN	T	*	1	1	0	1
Lissocarpaceae							
Lissocarpa guianensis Gleason	BARABARA, Charcoal	T		0	0	0	1
Loganiaceae							
Antonia ovata Pohl	INYAK	T		1	0	0	0
Strychnos bredemeyeri (Schult.) Sprague & Sandw.	DEVILDOER, Woody hook rough bark	L (W)	*	1	1	0	1
Strychnos diaboli Sandw. cf.	DEVILDOER, Woody hook, brown hairy	L (W)		0	1	0	1
Strychnos erichsonii M.R. Schomb.	DEVILDOER, Woody hook, glabrous	L (W)		0	1	0	0
Strychnos hirsuta Spruce ex Benth.	DEVILDOER, Woody Hook, hairy	L (W)	*	1	0	1	1
Strychnos melinoniana Baill.	DEVILDOER, Woody hook	L (W)	*	1	0	1	1
Strychnos subcordata Spruce	DEVILDOER, Woody Hook, small leaf	L (W)		0	1	0	0
Malpighiaceae							
Banisteriopsis martiniana (A. Juss.) Cuatrec. var. martiniana	Mabudehi	L (W)		0	1	0	0
Byrsonima aerugo Sagot	ARIKADAKO	T		0	0	1	1
Byrsonima stipulacea A. Juss.	KANOABALLI	T	*	1	0	0	0
Heteropterys cristata Benth. cf.		L (V)		1	0	0	0
Heteropterys multiflora (DC.) Hochr.		L (W)	*	1	1	1	1
Hiraea adenophora Sandw. cf.		L (W)		1	0	0	0
Hiraea affinis Miq.		L (W)		1	0	0	0
Hiraea fagifolia (DC.) A. Juss.		L (W)		1	0	0	0
Mascagnia sepium (A. Juss.) Griseb.		L (V)		1	1	0	0
Mezia includens (Benth.) Cuatrec.		L (W)	*	1	0	1	0
Stigmaphyllon sinuatum (DC.) A. Juss		L (V)		0	0	1	0
Tetrapterys calophylla A. Juss.		L (W)		1	0	0	0
Tetrapterys crispa A. Juss.		L (W)		1	0	0	0
Tetrapterys sp.		L (W)		1	0	0	0
sp.		L (W)		1	0	1	1
Marantaceae							
Ischnosiphon foliosus Gleason	MUKRU	U (H)		1	0	1	1
Ischnosiphon gracilis (Rudge) Körn.	MUKRU	U (H)		1	1	1	1
Ischnosiphon obliquus (Rudge) Körn.	MUKRU	U (H)		1	0	1	0

Species	Local name	Growth form >	dbh 10 cm	Pib	War	Res	2km
Ischnosiphon puberulus Loes. var. scaber (Peterson) L. Andersson	MUKRU	U (H)		1	1	1	1
Marcgraviaceae							
Marcgravia parviflora Rich. ex Wittm.	Aruakabo	HE		0	0	1	1
Norantea guianensis Aubl.	KARAKARA	HE		1	1	1	1
Souroubea guianensis Aubl. subsp. guianensis	KARAKARA	HE	*	0	0	1	0
Melastomataceae							
Bellucia grossularioides (L.) Triana	Sakwasepere, Mess apple	T		0	0	0	1
Clidemia conglomerata DC.		U (S)		1	1	0	1
Clidemia sp.		U (S)		1	0	0	0
Miconia argyrophylla DC. subsp. argyrophylla	WARAIA	T		1	1	1	1
Miconia argyrophylla DC. subsp. gracilis Wurdack	WARAIA	T		0	1	0	0
Miconia gratissima Benth. ex Triana	WARAIA	T		0	0	0	1
Miconia hypoleuca (Benth.) Triana	WAKARADAN	T		0	1	0	0
Miconia lepidota DC.	WARAIA	T		0	0	1	0
Miconia punctata (Dresr.) D. Don	WARAIA	T		1	1	1	1
Miconia sp.	WARAIA	T		0	0	0	1
Mouriri sp.	MAMURIBALLI	T	*	1	1	0	1
Tococa aristata Benth.	HURUEREROKO	U (S)		1	1	1	0
sp1		U (S)		1	1	1	1
sp2		U (S)		1	0	0	0
Meliaceae							
Carapa guianensis Aubl.	CRABWOOD	T	*	1	0	1	0
Trichilia sp.	YURIBALLI	T		0	0	0	1
sp.		T		0	0	0	1
Menispermaceae							
Abuta barbata Miers	DEVILDOER, Round leaf	L (W)	*	1	1	0	0
Abuta bullata Moldenke	DEVILDOER, Round leaf	L (W)		0	1	1	0
Abuta imene Eichler	DEVILDOER	L (W)		1	1	1	1
Abuta rufescens Aubl.	DEVILDOER, Hairy leaf	L (W)	*	1	0	1	1
Abuta sandwithiana Krukoff & Barneby	DEVILDOER, Small leaf	L (W)		1	1	1	1
Anomospermum grandifolium Eichler	DEVILDOER, Rough bark	L (W)	*	1	1	1	1
Curarea candicans (Rich.) Barneby & Krukoff	DEVILDOER, Flat bark	L (W)	*	1	1	1	1
Telitoxicum krukovii Moldenke	DEVILDOER, Climbing baromalli	L (W)	*	1	1	1	1
Telitoxicum minutiflorum (Diels) Moldenke	DEVILDOER	L (W)		0	0	1	1
Monimiaceae							
Siparuna decipiens (Tul.) A. DC.	IREK	T		0	0	1	0

Species	Local name	Growth form >	dbh 10 cm	Pib	War	Res	2km
Moraceae							
Brosimum guianense (Aubl.) Huber	TIBOKUSHI	T	*	1	0	1	0
Brosimum rubescens Taub.	DUKALIBALLI	T	*	1	0	1	0
Ficus guianensis Desv. cf.	FIGHE	*	1	0	1	1	
Helicostylis tomentosa (Poepp. & Endl.) Rusby	ITURI-ISHI-LOKODO	T	*	1	0	1	1
Pourouma guianensis Aubl. subsp. guianensis	BURUMA	T	*	1	0	0	0
Trymatococcus paraensis Ducke	PASTURE TREE	T	*	1	0	0	1
Musaceae							
Heliconia acuminata L.C. Rich.		U (H)		0	0	1	0
Myristicaceae							
Iryanthera sagotiana (Benth.) Warb.	KIRIKAUA	T		0	0	1	1
Virola michelii Heckel	DALLI, Hill	T	*	0	1	0	1
Virola sebifera Aubl.	DALLI, Man	T		1	0	0	0
Myrsinaceae							
Cybianthus sp.		T		0	1	1	1
Myrtaceae							
Calycolpus goetheanus (Mart. ex DC.) O. Berg	KAKIRIO, WILD GUAVA	T	*	0	1	1	1
Calyptranthes forsteri O. Berg	KAKI	T		1	0	0	0
Eugenia coffeifolia DC.	UNKNOWN C	T		1	1	1	1
Eugenia conjuncta Amsh.		T		1	0	0	0
Eugenia patrisii Vahl	HICHA	T		0	0	0	1
Eugenia sp.	TURTLE CHERRY	T	*	0	0	0	1
Myrcia sp.	IBIBANARO	T		1	0	1	1
Nyctaginaceae							
Neea sp.	MAMUDAN	T		1	0	0	0
Olacaceae							
Chaunochiton kappleri (Sagot ex Engl.) Ducke	HIWARADAN	T	*	1	0	1	0
Dulacia guianensis (Engl.) Kuntze cf.	HISHIRUDAN	T	*	1	1	0	0
Heisteria cauliflora Sm.	MAKARASALI	T		1	0	1	1
Maburea trinervis Maas	UNKNOWN Z	T		0	1	1	1
Minquartia guianensis Aubl.	WANANIA	T	*	1	0	0	0
Orchidaceae							
Vanilla cristato-callosa Hoehne cf.	Vanilla kamaye	L (V)		1	0	1	0
Vanilla sp.	Vanilla kamaye	L (V)		1	0	0	1
Palmae							
Astrocaryum gynacanthum Mart.	Urishi	T		1	0	0	0
Bactris balanophora Spruce		U (P)		1	1	1	1
Bactris humilis (Wallace) Burret	Yuruwe	U (P)		1	1	1	1
Bactris oligoclada Burret	Kidalebanaro	U (P)		1	1	1	1
Desmoncus macroacanthos Mart.	Kamwari, Hold-me-back	L (V)		0	0	0	0

Species	Local name	Growth form >	dbh 10 cm	Pib	War	Res	2km
Geonoma maxima (Poit.) Kunth var. Ambigua (Spruce) Henderson	Dhalebana	U (P)		1	1	1	1
Jessenia bataua (Mart.) Burret subsp. oligocarpa (Griseb. & H. Wendl.) Balick	TURU	T		1	1	0	1
Oenocarpus bacaba Mart.	LU	T	*	1	1	1	1
Passifloraceae							
Dilkea wallisii Mast. cf.		L (V)		1	0	0	0
Passiflora auriculata Kunth	PASSION FLOWER	L (V)		0	1	0	0
Passiflora cirrhiflora A. Juss.	PASSION FLOWER	L (V)		1	1	0	1
Passiflora coccinea Aubl.	PASSION FLOWER	L (V)		0	0	1	0
Passiflora fuchsiiflora Hemsl.	PASSION FLOWER	L (W)		0	0	0	1
Passiflora garckei Mast.	PASSION FLOWER	L (V)		0	1	0	0
Passiflora glandulosa Cav.	PASSION FLOWER	L (W)		0	0	1	1
Passiflora misera Kunth	PASSION FLOWER	L (V)		1	1	1	1
Passiflora vespertilio L.	PASSION FLOWER	L (V)		1	1	0	1
Passiflora sp1 (prob. spec. nov.)	PASSION FLOWER	L (V)		1	0	0	0
Piperaceae							
Piper hostmannianum (Miq.) C. DC.	Warakabakoro	L (V)		0	0	1	0
Polygalaceae							
Moutabea guianensis Aubl.	JULIAN ROPE	L (W)	*	0	1	1	1
Securidaca spinifex Sandw.		L (W)		1	1	1	1
Polygonaceae							
Coccoloba lucidula Benth. cf	Masari	L (W)		1	0	1	1
Coccoloba marginata Benth. cf.	Masari	L (W)		1	0	0	0
Coccoloba parimensis Benth.	Masari	L (W)		1	0	0	0
Coccoloba sp.	Masari	L (W)		1	0	0	1
Pteridophyta							
Lindsaea divaricata Klotzsch		U (H)		1	0	1	0
Lindsaea lancea (L.) Bedd. var. lancea		U (H)		1	0	0	1
Lygodium volubile Sw.		L (V)		1	1	1	1
Schizeae fluminensis Miers ex Sturm		U (H)		1	0	0	1
Trichomanes pinnatum Hedw.		U (H)		1	0	1	0
Triplophyllum funestrum (Kunze) Holttum		U (H)		1	0	0	0
Quiinaceae							
Quiina guianensis Aubl.	OKOKONSHI, Small leaf	T		0	1	1	1
Quiina indigofera Sandw.	OKOKONSHI, Broad leaf	T		1	1	1	1
Quiina obovata Tul. cf.	OKOKONSHI, Stelt roots	T		1	0	0	1
Quiina spp.	OKOKONSHI	T	*	0	0	0	1

Species	Local name	Growth form >	dbh 10 cm	Pib	War	Res	2km
Rapateaceae							
Rapatea paludosa Aubl. var. paludosa		U (H)		1	0	0	0
Rhizophoraceae							
Cassipourea lasiocalyx Alston		T		0	0	1	1
Rubiaceae							
Amaioua guianensis Aubl. var. guianensis	KOMARAMARABALLI	T		1	1	1	1
Duroia eriopila L.f. var. eriopila	KOMARAMARA	T		1	1	1	1
Faramea quadricostata Bremek.		U (S)		1	1	1	1
Isertsia hypoleuca Benth.	KAMADAN	T		0	1	1	1
Malanea macrophylla Bartl. ex Griseb. var. macrophylla		L (W)		1	0	0	0
Malanea sp.		L (V)		1	0	1	1
Palicourea crocea (Sw.) Roem. & Schult.		U (S)		1	1	1	1
Palicourea guianensis Aubl. subsp. occidentalis	BUCKWOOD	T		1	0	0	0
Psychotria apoda Steyerm.	Koyarakushi	U (S)		1	1	1	1
Psychotria capitata Ruiz & Pav. subsp. inundata (Benth.) Steyerm.		U (S)		1	1	1	1
Psychotria hoffmannseggiana (Willd. ex Roem & Schult. Muell. Arg.		U (H)		1	0	0	1
Psychotria puberulenta Steyerm.		U (S)		1	1	1	1
Psychotria sp.		T		1	0	0	0
Randia asperifolia (Sandw.) Sandw.		L (W)		0	0	1	0
Sabicea surinamensis Bremek. cf.		L (V)		1	0	0	0
Uncaria guianensis (Aubl.) J.F. Gmel.		L (W)		1	0	1	0
Rutaceae							
Zanthoxylum apiculata (Sandw.) Waterman	SADA	T		1	0	0	1
Sapindaceae							
Cupania hirsuta Radlk.	KULISHIRI, Black	T		1	0	0	0
Cupania scrobiculata Rich. cf.	KULISHIRI, White	T		1	0	0	0
Cupania sp.	KULISHIRI	T		0	1	1	1
Matayba oligandra Sandw. cf.	KULISHIRI, Common	T	*	1	0	0	0
Matayba sp	KULISHIRI, Like	T		0	1	1	1
Paullinia capreolata (Aubl.) Radlk.		L (W)		1	1	0	1
Paullinia ingaefolia Rich.		L (W)		1	0	0	0
Paullinia pachycarpa Radlk.		L (W)		1	0	1	1
Paullinia sp1		L (V)		1	0	0	0
Serjania paucidentata DC.	Hebechi-abo	L (V)		0	0	0	1
Talisia elephantipes Sandw.	WILD GENNIP	T		0	0	1	0
Talisia squarrosa Radlk.	MOROBALLI	T	*	1	1	1	1
Sapotaceae							
Chrysophyllum pomiferum (Eyma) T.D. Penn.	PARIPIBALLI	T		1	0	0	1

Species	Local name	Growth form >	dbh 10 cm	Pib	War	Res	2km
Chrysophyllum sanguinolentum (Pierre) Baehni	BARATABALLI	T	*	1	1	1	1
Ecclinusa guianensis Eyma cf.	BARTABALLI, Like	T	*	0	1	1	1
Manilkara bidentata (A. DC.) Chev. subsp. bidentata	BULLET WOOD	T		0	0	0	1
Micropholis venulosa (Mart. & Eichl.) Pierre	KUDIBIUSHI	T	*	1	0	1	1
Pouteria caimito (Ruiz & Pav.) Radlk.	ASEPOKOBALLI, Fine Leaf	T	*	1	0	1	1
Pouteria cladantha Sandw. cf..	AIOMORAKUSHI	T	*	1	0	0	1
Pouteria coriacea (Pierre) Pierre	MORABALLI	T	*	1	0	1	0
Pouteria egregia Sandw.	KOKORITIBALLI, Fine Leaf	T		1	0	0	1
Pouteria filipes Eyma cf.	KAMAHORA, Fine Leaf	T		1	1	1	1
Pouteria guianensis Aubl.	ASEPOKO	T	*	1	1	1	1
Pouteria reticulata (Engl.) Eyma subsp. surinamensis T.D. Penn.	KOKORITIBALLI, Common	T	*	1	1	1	1
Pouteria speciosa (Ducke) Baehni	SUYA	T	*	1	0	0	1
Pouteria trigonosperma Eyma	KAMAHORA, Medium leaf	T	*	0	1	1	1
Pouteria venosa (Mart.) Baehni subsp. amazonica T.D. Penn. cf.	KAMAHORA, Broad Leaf	T		1	0	0	0
Pouteria sp1	KOKORITIBALLI, Like	T		0	1	0	0
Pouteria sp2		T		0	1	0	0
Pouteria sp3	KAMAHORA	T		0	0	1	0
Simaroubaceae							
Picramnia guianensis (Aubl.) Jansen-Jacobs	HIKURIBIANDA	T		1	1	0	0
Simaba guianensis Aubl. subsp. guianensis		T		0	1	0	1
Simaba multiflora Adr. Juss.	HACHIBALLI	T	*	1	1	1	1
Simarouba amara Aubl.	SIMARUPA	T	*	1	0	1	0
Smilacaceae							
Smilax cumanensis Willd.		L (V)		0	0	1	0
Smilax poeppigii Kunth		L (V)		0	0	1	1
Smilax santaremensis A. DC.	Dorokwaropimpla	L (V)		1	0	1	1
Smilax syphilitica Willd.	Dorokwaroyuruwan	L (V)		1	1	0	1
Solanaceae							
Markea coccinea Rich.		HE		1	0	1	0
Markea sessiliflora Ducke		HE		0	0	1	0
Solanum adhaerens Roem. & Schult.		L (V)		0	0	1	0
Sterculiaceae							
Sterculia pruriens (Aubl.) K. Schum. var. ?	MAHO, Smooth Leaf	T	*	1	0	0	1

Species	Local name	Growth form	dbh > 10 cm	Pib	War	Res	2km
Sterculia rugosa R. Brown	MAHO, Rough Leaf	T	*	1	0	0	1
Tiliaceae							
Apeiba petoumo Aubl.	DURU	T	*	1	0	1	1
Verbenaceae							
Petrea macrostachya Benth.	SANDPAPER VINE	L (W)		1	0	0	1
Petrea volubilis L.	SANDPAPER VINE	L (W)		1	0	0	0
Violaceae							
Paypayrola guianensis Aubl.	ADEBORO, Like (with smell)	T		?	0	1	1
Paypayrola longifolia Tul.	ADEBORO	T		1	1	0	0
Vochysiaceae							
Ruizterania albiflora (Warm.) Marcano-Berti	MUNIRIDAN	T	*	1	0	0	0
Vochysia surinamensis Stafleu var. surinamensis	ITEBALLI	T	*	1	0	0	1
Zingiberaceae							
Renealmia orinocensis Rusby	INK-BERRY	U (H)		1	1	0	1
Family unknown							
indetT1	HURIBALLI	T		0	0	0	1
indet T spp.	UNKNOWN	T	*	1	1	1	1
undetL1		L (W)		0	0	0	1
undetL2		L (W)		1	0	0	0
undetL3		L (W)		1	0	0	0
undetL4		L (W)		0	0	1	0

Appendix 4.2
Species richness of other mixed tropical rain forest on terra firme in South America.

Source	Location & Forest type	Sample size & shape	dbh min		Number of plants	Fam	Genera	Species
Ek & van der Hout*,	Mabura Hill, Guyana	15 ha	10 cm	**min**	170			29
this study	Greenheart forest	15 of 100 x 100 m		**max**	248			59
Davis & Richards 1934	Moraballi Cr., Guyana	1.5 ha	10 cm		644			91
	terra firme (mixed)	400 ft x 400 ft						
Davis & Richards 1934	Moraballi Cr., Guyana	1.5 ha	10 cm		919			74
	terra firme (wallaba)	400 ft x 400 ft						
Cain *et al.* 1956[a]	Belém, Brazil	2 ha	10 cm		897	39	100	153
	terra firme	20 of 10 x 100 m						
Campbell *et al.* 1986*	Rio Xingu, Brazil	3 ha	10 cm		1420	39	127	265
	terra firme	10 x 3000 m						
Gentry 1985*	Manu, Peru	1 ha	10 cm		673	43		210
	terra firme	100 x 100 m						
Gentry 1988*	Yanomomo, Peru	1 ha	10 cm		605			283
	terra firme	10 x 1000 m						
Korning & Balslev 1994	Cuyabeno, Ecuador	1 ha	10 cm		697			313
	terra firme							
Hartshorn 1980*	Manu, Peru	1 ha	10 cm		584			153
	terra firme	no layout						
Pires 1966	Breves, Brazil	1 ha	10 cm		516	36		157
	terra firme	10 x 1000 m						
Prance *et al.* 1976[a]	Manaus, Brazil	1 ha	15 cm		350	43	115	179
	terra firme	irregular						
Rodrigues 1963[a]	Amapá, Brazil	1.1 ha	15 cm		347	84	63	36
	terra firme, ridge top	11 of 10 x 100 m						
Comiskey *et al.* 1994	Kwakwani, Guyana	1 ha	10 cm		504	35	65	85
	mixed forest	100 x 100 m						
Comiskey *et al.* 1994	Kwakwani, Guyana	1 ha	10 cm		493	29	49	59
	Greenheart forest	100 x 100 m						
Johnston & Gillman 1995	Kurupukari, Guyana	1 ha	10 cm		477	23		67
	Greenheart forest	100 x 100 m						
Johnston & Gillman 1995	Kurupukari, Guyana	1 ha	10 cm		357	23		64
	Mora forest	100 x 100 m						
Johnston & Gillman 1995	Kurupukari, Guyana	1 ha	10 cm		459	23		71
	Mixed forest	100 x 100 m						
Johnston & Gillman 1995	Kurupukari, Guyana	1 ha	10 cm		742	23		50
	Mixed forest	100 x 100 m						
Ek & van der Hout[b]	Mabura Hill, Guyana	15 ha	none	**min**				118
this study	Greenheart forest	15 of 100 x 100 m		**max**				175
Duivenvoorden 1994[b]	Araracuara, Colombia	1 ha	none			112	369	1233
		10 of 0.1 ha,						
		irregular shaped						
Bordenave 1996[b]	Sinnemary, French Guiana	0.5 ha	none			55	119	185
	terra firme	5 of 10 x 10 m						
Bordenave 1996[b]	St. George, Fr. Guiana	0.5 ha	none			64	135	197
	terra firme	5 of 10 x 10 m						

[a]From Campbell et al. 1986.; *: including lianas, [b]: including lianas, herbs, shrubs, and small palms. More examples in Richards (1996).

CHAPTER 5

The effects of logging on the liana vegetation in Greenheart dominated tropical rain forest in Guyana

Renske C. Ek, Cornel A. van der Kooij & Juul A.C. van Dam

Abstract

Patterns of variations in composition and development of liana vegetation over space and time as a result of logging are described for Greenheart (*Chlorocardium rodiei*) dominated mixed rain forest near Mabura Hill, Guyana. In order to study the relation between the composition of the liana community (in species richness, diversity and abundance), and the presence of different logging-related damage zones, the number of stumps of harvested trees, and the size and location of gaps and skid trails were measured.

Logging damage, or logging induced changed environment, leads to clear patterns of liana diversity. High liana species richness occurred both in logged and unlogged forest. Within the same site, however, species richness and abundance increased with logging. The number of all liana species after logging was positively correlated with the number of liana species before logging (previous occupation and opportunities of seed dispersal), and negatively correlated with the amount of logging damage, expressed in number of harvested trees and skid trail area. The number of newly established liana species after logging was negatively correlated with the amount of skid trail area. After logging, especially the number of common species increased, compared to rare species. Unlogged plots have a different liana vegetation composition, with lianas that prefer or can endure lower light levels. In the logged sites species become dominant with preferences for higher light levels.

Based on the behavior of individual species four different functional liana groups could be defined, each with its own response to logging. The behavior of individual species showed that although lianas are mostly claimed to be pioneer species, some clearly declined in numbers or disappeared completely as a result to logging, some reacted indifferently to logging and others proliferated after logging. The indifferent species could be furthermore divided in a group of species with showed clear preferences for higher light levels and a group with preferences for lower light levels.

Turnover of species and individuals for two plots enumerated before and two years after logging showed that especially the area outside gaps had positive correlations with number of species and individuals. These outer gap zones have increased light levels but no surface disruption and show favorable conditions for liana seedling establishment shortly after logging, both for species already present before the logging as for newly arriving species.

Introduction

Logging involves the felling and extraction of trees. Logging will lead to opening up the canopy, creating gaps in the broad sense (*e.g.* Connell & Lowman 1989; Ter Steege *et al.* 1996). The increased light availability in and around gaps is believed to be the most important factor controlling seedling establishment and performance in gaps (*e.g.* Clark 1990; Welden *et al.* 1991; Zagt 1997; Sterck 1997). The light conditions within gaps (Clark & Clark 1990) and in their immediate surroundings (Popma *et al.* 1988; Lieberman *et al.* 1989; Van der Meer & Bonger 1996a) may vary considerably as a result from the gap size, form and consequent position towards the prevailing course of the sun (Ter Steege 1993). Skidder activity (mechanized extraction) inflicts damage to the remaining stand (Bruijnzeel & Critchley 1994; Ter Steege *et al.* 1996), uprooting of the soil and soil compaction. On skid trails the hydrological characteristics of the soil may change (Jetten 1994), and due to the absence of sufficient living roots leaching of nutrients (mainly cations), soil acidification and raised Aluminum levels may occur (Brouwer 1996).

Regenerating forest areas are often particularly rich in lianas due to their strong resprouting abilities (Putz 1984; Appanah & Putz. 1984), and they are capable of quick recovery after stem damage (Fisher & Ewers 1992). In general, high density of lianas can be interpreted as biological markers of forest disturbance or successional structural crises (Hegarty & Caballé 1992). In relation to logging, the connecting ability of lianas (Gentry 1992) can result in that, at any event forming a gap, more trees collapse and the resulting gaps are larger (Hammond & Brown 1991; Van der Hout 1996). Lianas can also influence the growth of tropical trees and their seedlings (Putz 1984; Putz *et al. 1984;* Clark & Clark 1990). In view of the above, foresters consider lianas as a nuisance and pre-harvest liana cutting is part of many silvicultural treatments (Fox 1968; Appanah & Putz 1984; Putz 1992; Mason 1996; Van der Hout 1996).

In this study the term 'lianas' is used as a synonym for several types of climbers, which includes woody climbing plants that rely on other plants for support (Putz 1984), woody hemiepiphytes, which rely on other woody plants for their support for only part of their life cycle (Benzing 1995), and herbaceous tendril climbing plants (Gentry & Dodson 1987).

Functional units, *i.e.* functionally similar groups of species whose members share characteristics of importance for determining forest structure and composition, may simplify the analysis of diversity of tropical rain forest. The use of those groups, or guilds, reveals general (and specific) patterns of responses to logging regardless of geographical, phenotypical or taxonomical differences and facilitates predictions about forest processes. The oldest known use of guilds in the context of lianas is from Schimper (1903). In this paper we use guilds in the context of functional liana groups that share the same response to logging and show the same preferences for site of establishment.

The aim of this study is to reveal how the composition and development of liana vegetation varies over space and time as a result of logging. We study the relation between the composition of the liana community (in species richness, abundance and functional diversity) and the presence of different forest areas, e.g. small gaps, large gaps and skid trails. Both plots in natural Greenheart dominated mixed forest, and plots that had been logged 2, 6 and 10 years before, are enumerated. Changes in diversity of lianas due to logging are evaluated in terms of species diversity and functional diversity.

Methods

Research sites

The study has been conducted on the timber concession of Demerara Timbers Ltd. in the vicinity of Mabura Hill (5°13'N, 58°48'W), as part of The Tropenbos-Guyana Programme. The major aim of the program is to reach a level of understanding of the regional lowland tropical rainforest ecosystems which can be used in the design of a sustainable forest management system that avoids the loss of proper hydrological functions, nutrient cycles, or timber tree regeneration, and retains a satisfactory level of plant and animal diversity. The presented research is part of a study on the effects of logging on the botanical diversity. More information on the Tropenbos-Guyana Programme and the study site can be found in Ter Steege *et al.* (1996).

Greenheart (*Chlorocardium rodiei*) is Guyana's major commercial timber species and all study locations are situated in 'Greenheart dominated mixed forest on brown sand'. The concession is 15 years old, during the years the logging activities migrated further south into the concession, consequently the sites situated in closer vicinity of Mabura Hill township are logged longer ago. In total 24 hectares of Greenheart forest were inventoried (Fig. 4.1); in 1993 and 1994 15 one-ha plots natural Greenheart forest in the West Pibiri compartment (this site hereafter will be referred to as 'Pibiri'), in 1995 two one-ha plots logged in 1985 and one unlogged referential one-ha plot at 2 km south of Mabura township ('2 km'), in 1995 two one-ha plots logged in 1989 and one referential one-ha plot in kilometer square no. 34 in Waraputa ('Waraputa') and near the Mabura Hill Forest Reserve ('Reserve'), respectively. In 1996 two plots, logged in 1994, were re-enumerated in the Pibiri site. In relation to this paper mainly the two re-enumerated Pibiri plots were used, both with pre-logging and post-logging liana vegetation composition, together with the older logged sites.

Data collection

All 24 one-ha plots were divided in 20 x 20 m subplots and enumerated with the following intensity: large lianas with a dbh≥10 cm for every 20 x 20 m subplot; adult lianas with a height≥2 m in 10 x 10 m subplots, one in every 20 x 20 m plot; liana saplings (50 cm≤height<2 m) in 5 x 5 m subplots, one in every 10 x 10 m subplot; liana seedlings (height<50 cm) in 2 x 2 m subplots, one in every 5 x 5 m subplot.

More details on the sampling method and plot lay-out can be found in Van der Hout (1996) and Alder & Synnott (1992).

Members of a clonal group with above-ground connections were regarded as one individual. Individuals growing in the plot, but rooting outside, were not included. For each sampled individual species name, phenology, coordinates within the plot, height, diameter, diameter of host and climbing method was determined.

For all species vegetative voucher collections were made and deposited in Utrecht (U).

Gap area, measured as a vertical projection of a canopy opening enclosed by the crown edges of the surrounding trees (Putz 1984), was estimated using an octagon method connecting eight gap edge points at 45° intervals (after Lawton & Putz 1988; Jans *et al.* 1993). For equations used to calculate coordinates within the plots and the eight gap edge points, see appendix 1. Afterwards, we first defined an extended borderzone, an uncertainty zone of 0.5 m inwards and outwards of the measured gap border (the 'gap edge'), and different gap zones in and around gaps using a GIS (see further, 'spatial analyses and forest zonation').

Skid trails were traced using signs of low secondary vegetation, damage to remaining large trees, and the presence of stumps of harvested trees. The skid trail area was estimated by measuring the position of trail edges in relation to the 20 x 20 m grid of the plots, combined with positions of points where the trails divided in two or more trails, or changed direction. An extended borderzone of 0.5 m inwards and outwards was calculated afterwards.

Data analyses

Spatial analyses and forest zonations

To be able to combine all spatial data a Geographical Information Systems was used (PC-Raster (Van Deursen & Wesseling 1993)). The GIS system was used to integrate the measurements of the species composition, their relative abundance and location, and the location, size and form of gaps and skid trails. Furthermore, the forest was classified and zonated in two different ways: 1) Damage zonation; and 2) Fixed distance zonation (Fig. 1).

1) Damage zonation (Fig. 1a): Within the damage zonation we distinguished between gaps, large gaps (≥ 300 m2), small gaps (< 300 m2), skid trails, overlapping gap and skid trail, outer gap (a zone till 15 m from gap edge) and forest interior (non gap or skid trail area).

2) Fixed distance zonation (Fig. 1b): Under the assumption that around gaps diffuse light levels would increase in all directions we defined gap zones with a fixed width of 5 m, all related to distance to the edge of the gap. These gapzones were: central gap zone (> 5 m from gap edge), inner gap edge zone (0 - 5 m from gap edge), first outer gap edge zone (0 - 5 m), second outer gap zone (5 - 10 m), third outer gap zone (10 - 15 m), fourth outer gap zone (15 - 20 m) and adjacent forest (interior, > 20 m). Note: Under the assumption that light levels around gaps were influenced by

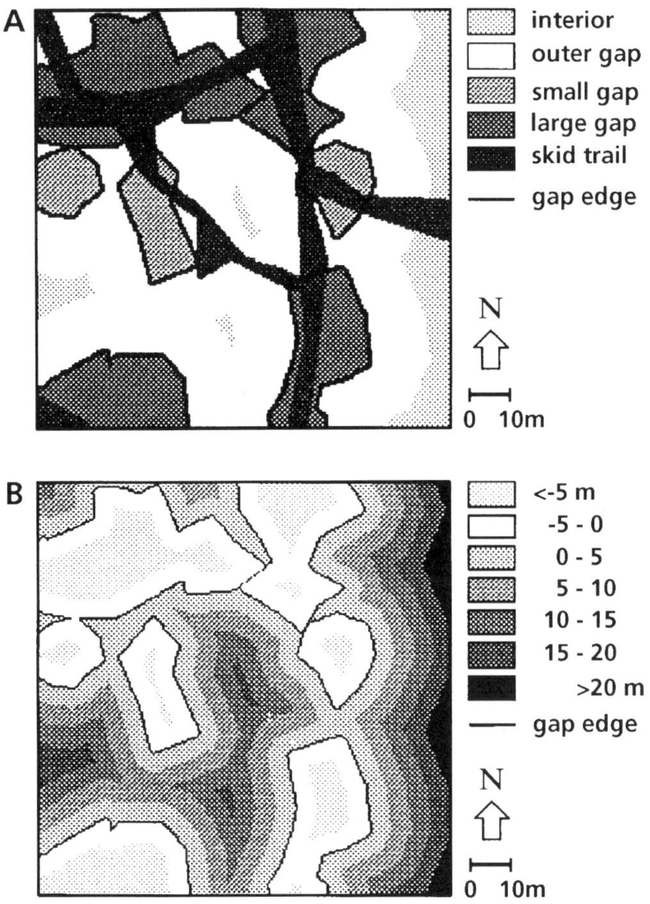

Figure 1
Examples for 1 one-ha at 'Waraputa', near Mabura Hill, Guyana, for two forest zonations, A, damage zonation and B, fixed distance zonation.

the size and position of the gap towards the prevailing course of the sun, combined with the height of the canopy, and that the attenuation of light with distance from the gap could be considered dependent on density of surrounding vegetation, we also calculated for all points in and around gaps the time of exposure to direct light through the gap as a percentage of full sunlight. These increased light levels over the plot after gap formation were scaled in a logarithmic scale. The results did not differ significantly from the fixed distance zonations in respect of interpretation, for this paper we only use the fixed distance zonation.

Please note that the precise definition of interior is different for the two zonations. For the different one-ha plots and subplots, calculations of area of all different forest zones were made. Using the positions of individuals in different subplots, we additionally used the GIS-systems to obtain the 'environmental data' set (see pattern analyses).

Table 5.1
Species richness of lianas in different diameter- and height classes, and their occurrence in subplots, together with the number of individuals, number of felled trees and effected forest areas for each one-ha plot in Greenheart dominated mixed forest, Mabura Hill, Guyana.

	Pibiri				*Waraputa*		
	logged in 1994				*logged in 1989-1990*		
	plot 2		*plot 7*				
Number of species:	UL	LOG	UL	LOG	UL	LOG1	LOG2
diam > 10 cm	11	6	9	8	3	3	0
height >2 m	49	63	54	64	39	44	45
height >0.5 m	21	51	21	47	12	18	16
height <0.5 m	9	23	8	28	7	8	8
total	56	78	47	73	39	47	46
*new after logging**		32		39		18	21
Occurrance:							
in 1 subplot only (rare)	18	22	18	21	11	14	14
>5 subplots (common)	18	30	19	25	9	13	17
ratio rare/common	1.00	0.73	0.95	0.84	1.22	1.08	0.82
number of individuals/ha	320	494	301	382	377	973	899
number of felled trees		16		16		16	25
total area (square meter):							
gaps	-	2583	-	4119	1655	3615	3569
skid trail	-	679	-	1215		1692	1613
total for gaps & skid trail	-	3350	-	4820		4225	4401
non gap or skid trail area	-	6650	-	5180	8345	5778	5599

Pattern analyses

The multi variate analyses we performed on the total data-set were mainly multiple linear regression and correlation coefficients (Pearson product moment).

Multiple linear regressions are performed on the level of one-ha plots. For the two Pibiri plots we knew exactly the numbers of species and individuals before and after logging. For the older logged sites we assumed the non logged one-ha plot as representative for the former unlogged status of the two logged plots, both in number of species and number of individuals.

All correlations are in general calculated on basis of 10 x 10 m subplots, although for the two Pibiri plots also 5 x 5 m subplots were used. By using subplots as unit instead of individuals we avoided spatial dependency of liana individuals of the same species. We assumed that for the older logged sites, logged in 1989-1990 and in 1985, lianas occurring in the 10 x 10 m subplots (diam < 10 cm, height > 2 m) would show patterns in distribution and diversity which reflect the environmental effects of

Table 5.1 (continued)

* : For the areas Waraputa, Reserve, and 2km, the unlogged plot is considered as representative for the pre-logging status of the logged plots.

	Reserve			*2 km*		
	logged in 1989-1990			*logged in 1985*		
Number of species:	UL	LOG1	LOG2	UL	LOG1	LOG2
diam > 10 cm	9	5	2	13	11	9
height > 2 m	34	58	56	53	58	65
height >0.5 m	28	29	25	19	22	28
height <0.5 m	17	13	10	21	14	16
total	43	62	60	69	59	68
*new after logging**		37	41		24	25
Occurrance:						
in 1 subplot only (rare)	15	22	16	30	13	20
>5 subplots (common)	12	24	20	13	30	25
ratio rare/common	1.25	0.92	0.80	2.31	0.43	0.80
number of individuals/ha	181	593	593	625	761	764
number of felled trees		20	21		29	22
total area (square meter):						
gaps	1066	2827	3949	1739	4112	3455
skid trail		1835	1732		2216	1632
total for gaps & skid trail		4007	4788		4758	4708
non gap or skid trail area	8934	5993	5213	8261	5242	5292

Table 5.2
Relation of a) number of liana species after logging, and b) newly established liana species after logging, with the liana species richness before logging and different damage variables, for 8 one-ha plots near Mabura Hill, Guyana.

	SP	SP-NEW
SP-OLD	0.88 *	
STUMPS	-0.77 *	
SKID	-0.88 *	-0.73 *
R-square	0.87 *	0.53 *

logging 6 and 10 years earlier. For the site logged in 1994, however, we assumed that the establishment after the logging would be reflected in changes in species composition and abundance in the smaller size classes (< 2 m height). Therefore, this smaller size class is also used in the analyses and the results from the older logged plots are compared with this newly established vegetation. The species richness and abundances of lianas and individual species were correlated with the amount of area of

Table 5.3
Correlation of number of liana individuals and number of liana species per subplot for two different forest zonations occurring in 11 one-ha plots in Greenheart dominated mixed forest, Mabura Hill, Guyana.

	All (N=275)		unlogged (N=75)		logged (N=200)	
	IND	SP	IND	SP	IND	SP
Damage zonation:						
GAP	0.24 **	0.25 **	ns	ns	0.21	0.24
large	0.22 **	0.23 **	ns	ns	0.20	0.22 **
small	ns	ns	ns	ns	ns	ns
SKID	0.20 **	0.22 **			0.15	0.17
GAP-SKID	0.26 **	0.20 **			0.21	ns
OUTERGAP	ns	ns	0.29	ns	-0.25 **	-0.27 **
INTERIOR	-0.37 **	-0.34 **	-0.38 **	-0.24	-0.19	-0.16
Fixed distance zonation:						
> 5m in	0.17	0.16	ns	ns	ns	ns
0-5m in	0.24 **	0.23 **	ns	ns	0.22	0.22
EDGE	0.20	0.20	ns	ns	ns	ns
0-5m out	0.14	0.20	0.32	0.23	ns	ns
5-10m out	ns	ns	ns	ns	ns	-0.14
10-15m out	ns	-0.17	ns	ns	ns	-0.20
15-20m out	-0.12	ns	ns	ns	ns	ns
> 20m out	-0.35 **	-0.32 **	-0.41 **	-0.31	-0.16	ns

specific forest zones. Correlation coefficients were tested with Student t tests. For the statistical analyses STATISTICA 4.5 (StatSoft, 1993) was used.

Results

Species diversity

In total we recorded 179 liana species, of which 167 (93.3%) could be identified to species level, 8 species (4.5%) to genus level and 4 species (2.2%) remain unidentified. For a list of all species and their authorities, together with their presence in the subplots at the different sites, see appendix 2.

The species richness of lianas in different diameter and height classes and their occurrence in the subplots is given in table 1. In general, logged plots had a higher species richness than unlogged plots. This is true for all sites for the height class of lianas > 2 m. For the total species richness over all size classes the unlogged plot of 2 km formed the exception. This plot was species rich both in the largest (diam > 10 cm), and the smallest (height < 0.5 m), size class, and furthermore it had the highest number of species that occur in one subplot only. The ratio rare species to common species calculated per one-ha plot (species occurring only in one subplot/ species occurring in more than 5 subplots), showed that this ratio was indeed the highest for the unlogged plot of 2 km (2.31). This ratio decreased with logging at all sites, meaning that on the level of one-ha the number of common species after logging increased more than the number of rare species.

To investigate whether or not species richness and abundance after logging could be regarded as dependent on the amount of damage as a result of logging, a multiple regression was performed for all logged plots (N=8) with the total number of harvested trees, number of dead standing trees, total gap area, total skid area, total effected area by skid trails and gaps, and non effected area (table 1). The total number of species after logging proved to be positively correlated with the number of species before logging, and negatively correlated with number of harvested trees and skid trail area (R-square= 0.87, $p < 0.05$) (table 2). None of those variables were separately significantly correlated (Pearson product moment correlation) with the number of species after logging (data not shown). The relation that explained most of the variation between the plots concerning the newly established species after logging had only one independent variable, total skid trail area (-) (R-square=0.53, $p<0.05$ (table 2). A significant relation with number of individuals per plot and the amount of damage due to logging was not found.

Diversity on the scale of subplots: the occurrence of specific forest zones

To investigate the correlation between species richness and abundance of lianas with the occurrence of different forest zones, we calculated the Pearson product-moment correlation for all subplots together (N=275), for all unlogged subplots (N=75) and all logged subplots (N=200). The results are summarized in table 3. Both damage zonation and fixed distance zonation showed, for all subplots together, significantly positively correlations with the more open and illuminated forest areas, both for number of species and individuals, *i.e.* gaps and in particular large gaps (t- test; $p<0.001$), skid trails ($p<0.001$) and overlapping gap and skid trail areas ($p<0.001$) within the damage zonation, and from the center of a gap till 5 m from the gap edge ($p<0.05$, except the first inner gap zone $p<0.001$) within the fixed distance zonation. Dividing the subplots in logged and unlogged ones showed that the correlations with the more open and illuminated forest areas are mainly found within the logged subplots. For the unlogged subplots only a negative correlation with the interior remains within the damage zonation, and especially the number of individuals showed a strong negative correlation ($p<0.001$). Within the fixed distance zonation, unlogged subplots showed higher correlations compared with logged subplots, with especially significantly less numbers of individuals and species per area in a zone more than 20 meters from the gap edge and with the first outer gap zone (till 5 meter out) as an important zone with significantly more individuals and species.

Evaluating the four different sites separately (table 4), we found that the Waraputa

Table 5.4
Correlation of number of liana species and number of individuals per subplot for two different forest zonations for four different sites in Greenheart dominated forest, near Mabura Hill, Guyana.

	Pibiri		*Waraputa*			
			unlogged		*logged in 1989-'90*	
			N=25		N=50	
	IND	SP	IND	SP	IND	SP
Damage zonation:						
GAP	ns	0.51	ns	ns	0.29	0.38
large	ns	ns	0.53	0.69 **	0.29	0.38
small	ns	ns	ns	ns	ns	ns
SKID	ns	ns			ns	ns
GAP-SKID	ns	ns			0.52 **	0.29
OUTERGAP	ns	ns	ns	ns	-0.46 **	-0.41
INTERIOR	ns	ns	ns	ns	ns	-0.31
Fixed distance zonation:						
> 5m in	ns	ns	0.48	0.47	0.29	0.30
0-5m in	ns	0.43	ns	ns	0.38	0.31
EDGE	ns	ns	ns	ns	0.33	0.40
0-5m out	ns	ns	ns	ns	ns	ns
5-10m out	ns	-0.28	ns	ns	-0.35	-0.30
10-15m out	ns	ns	ns	ns	-0.36	-0.51 **
15-20m out	0.29	ns	ns	ns	ns	-0.32
> 20m out	ns	ns	ns	ns	ns	ns

site showed the strongest correlations between number of individuals and number of species with the two different forest zonations. All sites showed their own specific responses to logging, dependent on the level of disturbance and total area of the different forest zones. Nevertheless, some general observations could be made. All older logged sites showed positive correlations between number of individuals and number of species with the damage classes gap, skid trails and overlapping gap and skid trail area, although not significantly for the Reserve site. In the most recently logged site, Pibiri, we only found positive correlations between the total gap area and the number of species. The first inner gap zone was the only zone significantly positively correlated with the number of species. The positive correlation between number of individuals and areas of higher light levels, as found in all other logged sites, were not present two years after logging. This could mean that two years after logging we still see mainly logging damage, and no other disappearance of species.

The forest zones mentioned above could have an influence on the abundance and preferences of individual species. To study this, we repeated the analyses on the level of all subplots, but this time with the total number of individuals per subplot for 50 species. These species were selected based on their frequencies in the four different

Table 5.4 (continued)

	Reserve				*2km*			
	unlogged		*logged in 1989-'90*		*unlogged*		*logged in 1985*	
	N=25		N=50		N=25		N=50	
	IND	SP	IND	SP	IND	SP	IND	SP
Damage zonation:								
GAP	ns	ns	ns	ns	ns	ns	0.35	0.35
large	ns	ns	ns	ns	ns	ns	ns	ns
small	ns	ns	ns	ns	ns	ns	0.29	ns
SKID			ns	ns			0.28	ns
GAP-SKID			ns	ns			ns	ns
OUTERGAP	ns	ns	ns	-0.34	ns	ns	ns	ns
INTERIOR	ns	ns	ns	ns	ns	ns	-0.4 *	-0.45
Fixed distance zonation:								
> 5m in	ns	ns	ns	ns	ns	ns	0.32	ns
0-5m in	ns	ns	ns	0.30	ns	ns	ns	0.30
EDGE	ns	ns	ns	ns	ns	ns	ns	ns
0-5m out	0.40	ns	ns	ns	ns	ns	ns	ns
5-10m out	ns	ns	ns	-0.31	ns	ns	ns	ns
10-15m out	ns	ns	ns	ns	ns	ns	ns	ns
15-20m out	0.41	0.49	ns	ns	ns	ns	-0.3	-0.36
> 20m out	ns	ns	ns	ns	ns	ns	-0.39	-0.39

sites. Those species that showed significant correlations with one of the forest zones are listed in table 5. Based on those analyses we first distinguished three groups of species: 1) a group of species correlating positively with either logging status of the forest, large gap, or one of the damage zones. Although the last four species listed in this group (table 5) did not correlate significant (or even positively) with logging (status=1), they did seem to be significantly favored by specific logging induced environmental circumstances. We concluded that these species proliferate after logging; 2) a group of indifferent species with no significant correlation with either status of the forest, gap size, or damage zone, and 3) a small group of species correlating negatively to logging, disappearing as a result of logging. This group was rather small due to the fact that most disappearing species are mainly site specific; these two species were the only ones abundant enough over the sites.

Species from the first group, *group 1,* that proliferate after logging, mainly correlated within the fixed distance zonation significantly positively with gap center and first inner gap zone ('>5 m in' and '<5 m in') and/or negatively with the most undisturbed zones ('>20 m out') with lowest increased light levels. Five species, however, showed no significant correlation at all with the fixed distance zonation,

Table 5.5
Correlations between number of individuals per liana species and occurrence of forest zones, near Mabura Hill, Guyana.

	Group	Corr	Mean		Damage zonation*				
			UL	LOG	GAP	SKID	GAP & SKID	OUTER GAP	INTERIOR
Pinzona coriacea	1	**0.31****	37.3	466.0	**0.19**	**0.19**	**0.37**	**-0.19**	**-0.24**
Passiflora spp.	1	**0.28****	2.7	87.0	ns	0.26	0.27	-0.12	-0.19
Maripa scandens	1	**0.24****	60.0	178.0	**0.18**	ns	ns	ns-	**-0.21**
Coccoloba parimensis	1	**0.13**	1.3	16.0	**0.14**	ns	ns	ns-	ns-
Anemopaegma parkeri	1	**0.16**	17.3	65.5	**0.14**	ns-	0.18	ns-	**-0.13**
Curarea candicans	1	**0.12**	0.0	12.5	ns	ns-	0.14	ns-	ns-
Mascagnia sepium	1	**0.12**	1.3	14.5	ns	ns-	0.14	ns-	ns-
Aristolochia daemoninioxia	1	**0.12**	1.3	12.5	**0.13**	ns-	ns-	ns	ns-
Tontelea coriacea	1	**0.19**	37.3	158.5	ns	**0.15**	ns-	ns-	ns-
Machaerium madeirense	1	**0.12**	13.3	39.5	ns	**0.12**	ns	ns	ns-
Rourea pubescens	1	**0.13**	60.0	107.5	ns	ns	ns	ns-	**-0.13**
Paullinia pachycarpa	1	**0.12**	0.0	6.0	ns	ns-	ns-	ns	ns-
Machaerium myrianthum	1	ns	2.7	8.0	**0.13**	ns-	**0.18**	**-0.14**	ns-
Strychnos bredemeyeri	1	ns-	2.7	2.0	**0.13**	ns-	**0.12**	ns-	ns-
Memora racemosa	1	ns	6.7	27.0	ns	ns-	ns-	ns-	ns-
Moutabea guianensis	1	ns-	32.0	28.0	ns-	**0.17**	ns-	ns-	ns
Lonchocarpus negrensis	2A	ns	32.0	41.0	ns	ns	ns	ns	ns-
Anomospermum grandifolium	2A	ns-	46.7	41.0	ns	ns	ns-	ns	ns-
Tetracera volubilis	2A	ns	14.7	13.5	ns	ns	ns	ns-	ns-
Petrea macrostachya	2A	ns	9.0	10.5	ns	ns	ns-	ns-	ns-
Doliocarpus major	2A	ns	4.0	5.0	ns	ns	ns	ns-	ns-
Connarus perrottetii	2B	ns	462.7	477.5	ns	ns-	ns-	ns	ns-
Forsteronia guyanensis	2B	ns	4.0	4.0	ns	ns-	ns	ns-	ns
Dalechampia olympiana	2B	ns	1.3	4.0	ns-	ns	ns-	ns	ns-
Pleonotoma albiflora	3	**-0.14**	6.7	1.0	ns-	ns-	ns-	ns-	ns
Memora moringifolia	3	**-0.14**	93.3	51.5	ns	ns-	ns-	ns	ns

Table 5.5 (continued)

	Gap Size		Fixed distance zonation							
	LARGE GAP	SMALL GAP	IN > 5m	IN 0-5m	EDGE	OUT 0-5m	OUT 5-10m	OUT 10-15m	OUT 15-20m	OUT > 20m
Pinzona coriacea	0.21**	ns	0.24	0.21	0.17	ns	-0.13	-0.15	-0.16	-0.18
Passiflora spp.	0.18**	ns-	0.17	ns	ns	ns	ns-	ns-	-0.12	-0.15
Maripa scandens	0.13	ns	ns	0.18	0.19	ns	ns-	ns-	ns-	-0.19
Coccoloba parimensis	0.12	ns-	0.14	ns	ns	ns-	ns-	ns-	ns-	ns-
Anemopaegma parkeri	ns	ns	ns	0.16	ns	ns	ns-	ns-	ns-	ns-
Curarea candicans	ns-	ns	ns	ns	ns	ns	ns-	ns-	ns-	ns-
Mascagnia sepium	ns-	ns-	0.15	ns	ns-	ns-	ns-	ns-	ns	ns-
Aristolochia daemoninioxia	ns	ns	ns	ns	ns	ns	ns-	ns-	ns-	ns-
Tontelea coriacea	ns	ns-	ns-	ns	ns	ns	ns	ns	ns-	ns-
Machaerium madeirense	ns	ns	ns	ns	ns	0.15	ns	ns-	ns-	-0.12
Rourea pubescens	0.13	ns	0.12	ns	ns-	ns	ns	ns-	ns-	ns-
Paullinia pachycarpa	ns	ns-	ns	ns	ns	ns	ns-	ns-	ns-	ns-
Machaerium myrianthum	0.17	ns-	0.20	ns	ns-	ns-	ns-	ns-	ns-	ns-
Strychnos bredemeyeri	ns	ns	0.16	ns	ns	ns	ns-	ns-	ns-	ns-
Memora racemosa	0.20**	ns-	0.13	ns	ns-	ns-	ns-	ns-	ns-	ns-
Moutabea guianensis	ns-	ns-	ns-	ns-	ns	ns	ns-	ns	ns	ns
Lonchocarpus negrensis	ns	ns	ns	ns	ns	ns	ns	ns-	ns-	ns-
Anomospermum grandifolium	ns	ns	ns-	0.15	ns-	ns-	ns-	ns	ns	ns-
Tetracera volubilis	ns	ns	ns-	0.14	0.16	0.14	ns-	ns-	ns-	ns-
Petrea macrostachya	ns	ns	ns-	ns-	ns	0.12	ns	ns-	-0.12	ns
Doliocarpus major	ns-	ns-	0.15	ns-	ns-	ns-	ns	ns	ns-	ns-
Connarus perrottetii	ns-	ns	ns-	ns	ns-	ns-	ns	0.12	0.13	-0.18
Forsteronia guyanensis	ns-	ns	ns	ns	ns-	ns-	-0.12	ns-	ns	ns
Dalechampia olympiana	ns-	ns	ns-	ns-	ns-	ns-	ns	0.12	ns	ns-
Pleonotoma albiflora	ns-	ns-	ns	ns-	ns-	ns-	ns-	ns	ns	ns
Memora moringifolia	ns	ns	ns-	ns	ns	ns	ns-	ns-	ns-	ns

Table 5.6
Turnover in number of liana species for two one-ha plots in Pibiri, near Mabura Hill, Guyana, before and after logging.

	Pibiri (logged in 1994)									
	Logged 1					Logged 2				
	total	diam	height			total	diam	height		
		>10cm	>2m	<2m	<50m		>10cm	>2m	<2m	<50cm
# species:										
before logging	**56**	11	49	21	9	**47**	9	54	21	8
after logging	**78**	6	63	51	23	**73**	8	64	47	28
same	**46**	5	27	13	9	**34**	8	35	14	7
new	**32**	1	36	38	14	**39**	0	29	33	21
dissapeared	**10**	6	22	8	0	**13**	1	19	7	1

	Shared (occurring in both)				
	total	diam	height		
		>10cm	>2m	<2m	<50m
# species:					
before logging	**28**	4	31	13	6
after logging	**45**	2	40	31	12
same	**24**				
new	**30**				
dissapeared	**13**				

four of those five species: *Curarea candicans, Aristolochia daemoninioxia, Tontelea coriacea* and *Moutabea guianensis,* correlated significantly with skid trail and/or overlapping skid trail and gap area and these forest zones are not accounted for in the fixed distance zonation. The fourth species, *Paullinia pachycarpa,* did correlate positively with the mentioned zones, but just under the level of significance.

Species from the second group seem to harbor two patterns and we divided them therefore in two different groups: *group 2A,* a group of species indifferent to logging, with positive correlations with the first four zones from the fixed distance zonation ('> 5m in' till '<5 m out') and/or negative correlation with the most undisturbed zones ('>20 m out'); and *group 2B,* a group of species indifferent to logging, with no such positive or negative correlations and/or with positive correlations with zonations further from the gap edge into the interior (mainly '<15 m out' and '<20 m out').

Species from the group that correlated significantly negative to logging, *group 3,* showed no significant correlations with the fixed distance zonation, but did seem to prefer specific light levels; *e.g. Memora moringifolia* preferred the somewhat higher increased light levels ('5 m in' till '5 m out') and only *Pleonotoma albiflora* seemed to

Table 7.
Total area for the different damage classes according to the damage zonation for two one-ha plots in Pibiri, near Mabura Hill, Guyana.

	Area (m^2)				
	GAP	SKID	GAP-SKID	OUTER GAP	INTERIOR
logged1	2671	588	121	5721	929
logged2	3606	425	790	4699	481

prefer the sites in the forest with lowest increased light levels, which were the most undisturbed remaining forest patches.

As a result we distinguished four different liana groups, each with its own response to logging. The underlined species were the two most abundant species for each group and were used as examples for their group.

Pibiri: composition of the liana vegetation before and after logging

Two plots in Pibiri were first enumerated in 1993, logged in 1994, and re-enumerated in 1996. The turnover in species composition is given in table 6. Although the number of species increased after logging, with 32 and 29 newly established species for the two plots, respectively, a number of species disappeared (10 and 16, respectively). Main decrease in species numbers occurred in the size category of lianas with a diameter less than 10 cm and a height more than 2 meters, decreasing with 22 and 19 species, while highest number of newly established species could be found in the height class between 50 cm and 2 meters. The lianas in the largest size category, with a diameter of more than 10 cm, only vanished as a result of logging (6 and 1 species). Only one new species emerged in this size class, due to one individual that just passed the size limit in between the two enumerations. The amount of shared species increased slightly for the two plots after logging; logged1 had 36 shared species out of 56 in total, which changed to 54 out of 78 species (64% to 69%) and logged2 had 36 out of 54 species, which changed to 54 out of 67 (67% to 80%). After logging the plots were more similar in species composition. Differences between the two plots (table 7) were a result from differences in effected area by gaps and skid trails, while both plots had a logging intensity of 16 trees/ha (Van der Hout 1996).

The influence of the different forest zonations on the establishment of new species and individuals is already evaluated in general (table 3). For Pibiri, however, it was possible to relate the species richness and number of individuals after logging with the species richness and number of individuals at exactly the same spot, before logging. In this analysis we included one more zone within the damage zonation, in the category overlapping gap and skid trail, we distinguished between the overlap with large gaps (≥ 300 m^2) or small gaps (<300 m^2). The differences between gap and skid trail area for the two plots resulted in differences in reaction of the liana vegetation. Still, some general observations could be made. Two years after logging the main increase in

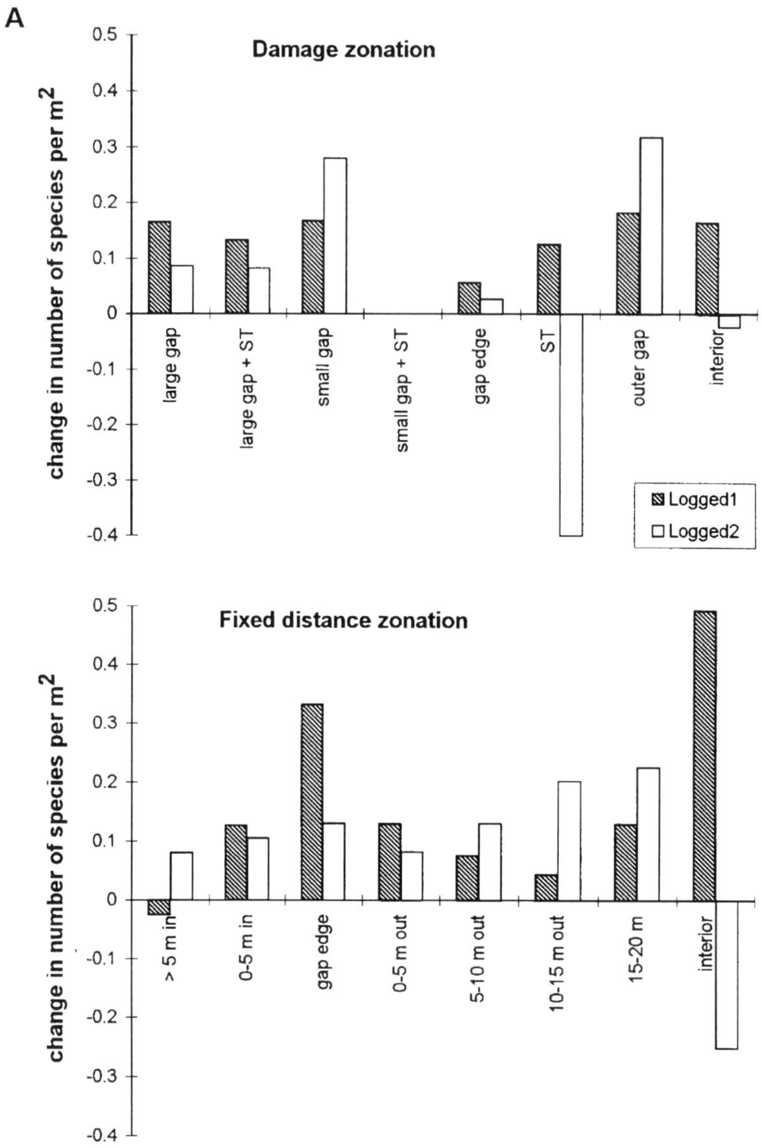

Figure 2
Relative changes in A) species richness per m^2 and B) number of individuals per m^2 for two one-ha plots in Pibiri, near Mabura Hill, Guyana, occurring in two different forest zonations before and after logging. Liana individuals used had a height between 50 cm and 2 meters.

B

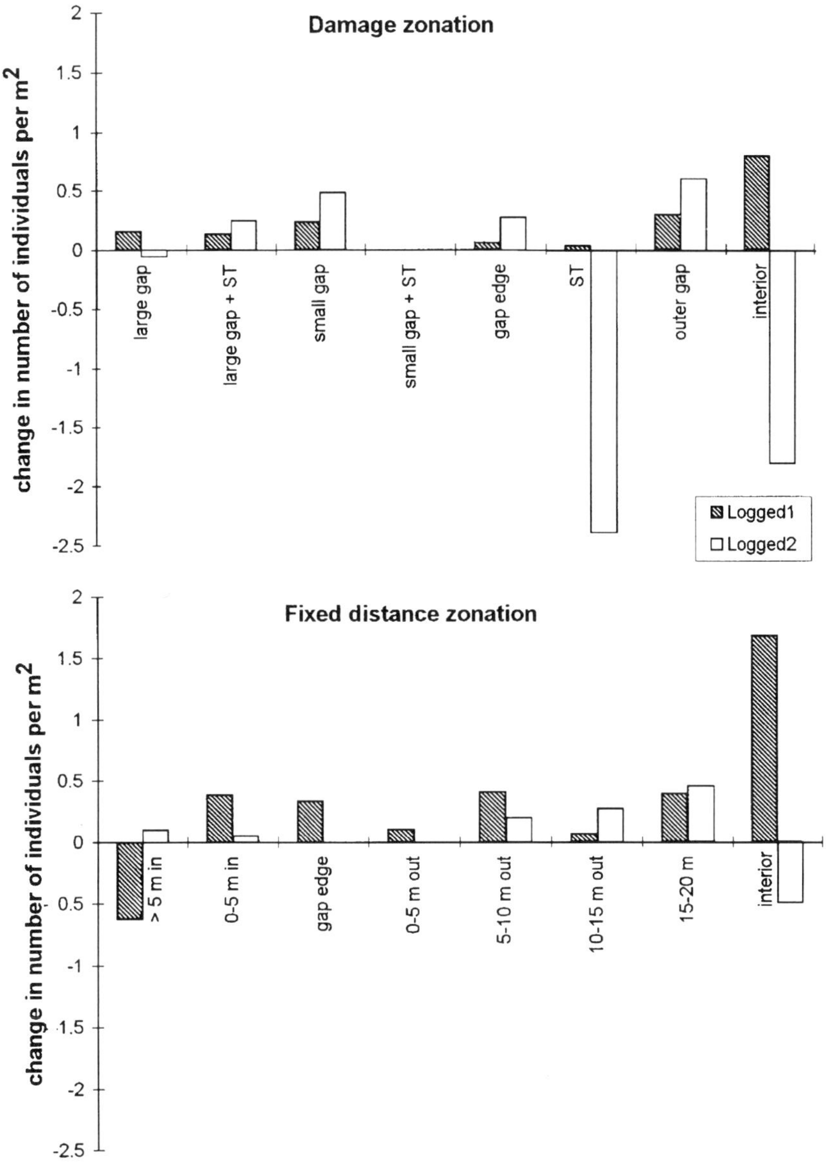
Damage zonation
change in number of individuals per m2
2
1.5
1
0.5
0
-0.5
-1
-1.5
-2
-2.5
large gap
large gap + ST
small gap
small gap + ST
gap edge
ST
outer gap
interior
Logged1
Logged2
Fixed distance zonation
change in number of individuals per m2
> 5 m in
0-5 m in
gap edge
0-5 m out
5-10 m out
10-15 m out
15-20 m
interior

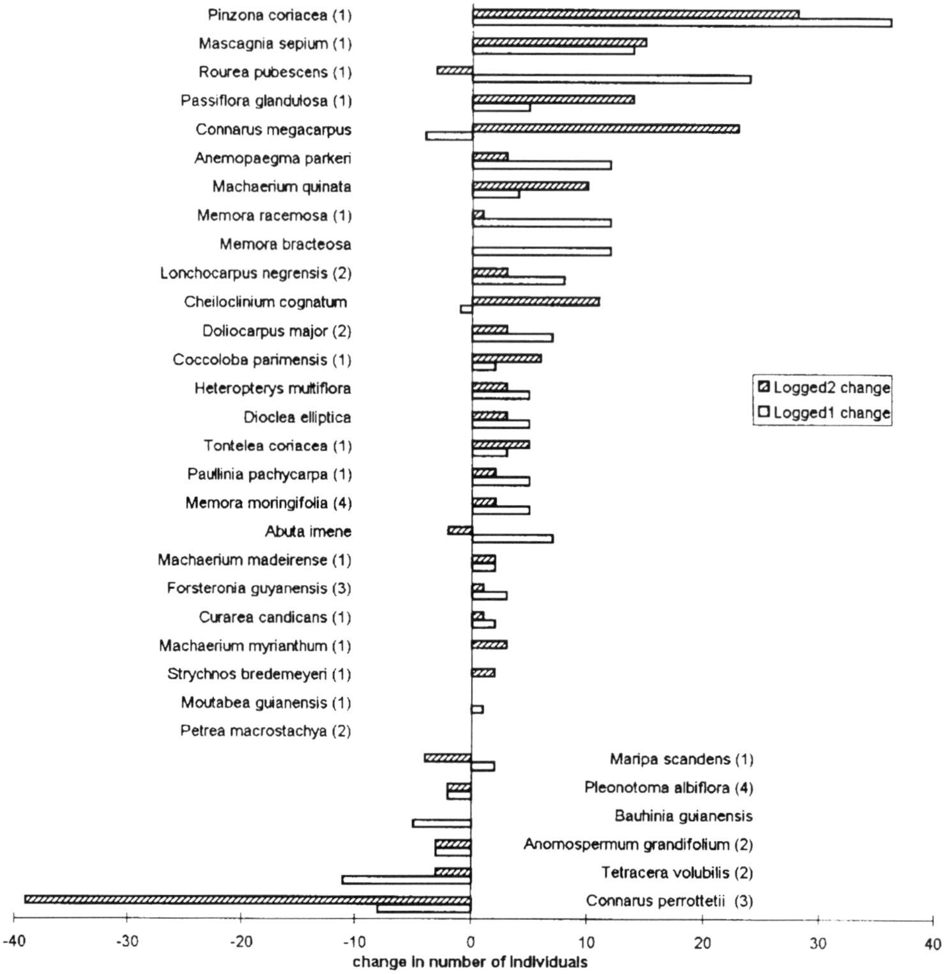

Figure 3

Change in liana sp[ecies abundance for two one-ha plots in Pibiri, near Mabura Hill, Guyana, before and after logging. Liana individuals used had a height between 50 cm and 2 meters. The numbers behind the species relate to the different groups resulting from table 5. If species do not have a number than they did not show significant correlations for all subplots together.

species numbers and number of individuals occurred in the outer gap zone and within small gaps (Fig. 2a&b). Decrease of species numbers and number of individuals was mainly found at skid trails and in zones with the extremes of light availability (gap center and interior) (Fig. 2a&b). Species exhibited a preference for both the areas around a gap edge, and the gap zones from 15 meters from the gaps, while for individuals especially the zones outside the gap were most important for establishment.

Species that showed highest increase of numbers of individuals two years after logging (in both plots > 10 individuals) (Fig. 3) are *Pinzona coriacea, Mascagnia sepium, Rourea pubescens* and *Passiflora glandulosa,* and all belonged to group 1, the group of species positively related to logging. In general all species from group 1 growing in Pibiri increased after logging, but *Maripa scandens* decreased in the plot with the largest logging affected area, logged2 (table 7). *Pleonotoma albiflora,* a member of group 3: declining species, did already decline in numbers of individuals shortly after logging, while *Memora moringifolia,* did seem to favor from the increased light levels that occur as a result of logging (table 5). Members of group 2A and 2B, species indifferent to logging, showed both increase and decrease in numbers of individuals, *Connarus perrottetii* (group 2B), *Teracera volubilis* and *Anomospermum grandifolium* (group 2A) all decreased in number of individuals, while the others increased.

Discussion and conclusions

Diversity

On the level of total botanical diversity of sites, plots, subplots, and forest zones, Ek & Van der Hout (chapter 4) describe that distance explained a large part of the variance in similarity of total species composition between, and within plots, when compared to other measured environmental variables, including logging. Still, we were able to find patterns of diversity of lianas clearly correlated with logging damage or logging induced changed environment.

High liana species richness could be found both in natural and logged forest, although in general, within the same site, logged forest had higher species numbers and number of individuals (table 1). On the level of one-ha, the number of species after logging was positively correlated with the number of species present before logging (previous occupation and opportunities of seed dispersal), and negatively correlated with the amount of logging damage, expressed in number of harvested trees and skid trail area (table 2). Number of newly established species was negatively correlated with skid trail area (table 2). Ek & Van der Hout (chapter 4), however, found for the newly established species after logging a positive correlation with overlapping gap and skid trail area and a negative correlation with the number of dead standing trees and total extent of the gap area. Their analysis is based on 6 sampling points, for the older logged sites. Our analysis is based on 8 sampling points, including the two re-enumerated Pibiri plots. In the two Pibiri plots overlapping gap and skid trail area is kept to a minimum, and hardly occurred. This shows that, as already suggested in chapter 4, the extent of overlapping gap and skid trail is important for changes in the species composition. The multiple regressions indicate that logging will always lead to an increase of liana species, as a consequence of creating new establishment sites. In general, with moderate logging intensities resulting in a mix of both disturbed and undisturbed forest patches, the more species rich the forest is before logging, the more species will have a change to establish themselves after logging. New, invading species were mostly found within heavily disturbed areas, *i.e.* the skid trail areas. After logging, especially the number of common species increased, compared to rare species (table 1). Higher logging intensities, *i.e.* increased numbers of harvested trees, will lead to an increased area that is heavily disturbed and decreased area of undisturbed forest. Consequently, more common species will appear, at the cost of rare species, and the species richness after logging will decrease. The fact that number of harvested trees performed better in the multiple regression than total gap area, could indicate that part of the original gap area was not recognized as such, especially in the sites logged longer ago. It was not possible to find a significant relationship between number of individuals and logging damage. The only relationship found was a negatively correlation with the number of species before logging together with a positively correlation with 'time after logging'. Although these results suggest that changes in species composition on the scale of a whole hectare occur shortly after logging and remain visible through time, while number of individuals changes over time, it should be noted that the site logged the longest ago, 2km, is also the most diverse site (in species richness and abundance). This influences all relationships concerning time after logging.

Older logged sites (table 1) have an increase in species richness mainly in the size category of lianas with a height of more than 2 meters, while for the site logged two years earlier an increase in species richness can be found in all size categories except for large lianas with a diam > 10 cm. These results confirm our assumptions. We assumed that for the older logged sites large lianas with a diam > 10 cm are expected to be remnants of the pre-logging liana vegetation; their presence could merely indicate that they did not die. Seedlings and saplings have a shorter history and could have established, as a result of the logging 6 and 10 years ago, under the newly formed post-logging vegetation. For the older logged sites, therefore, we assumed the size category of lianas with a height > 2m as representative for the changes in species composition and abundance as a result of logging. For the Pibiri site, logged two years earlier, the changed environment after logging is expected to have favored seedlings and saplings already present before the logging and resprouts of fallen down and damaged lianas and that the effects as such are visible in the size category of lianas with a height > 2m (after Putz 1984). However, changes in species composition and abundance in the smaller size classes (<2 m height) are expected to reflect establishment after the logging.

Preferential establishment sites of lianas

Gaps

Lianas are generally considered to be successional plants, colonizing open sites (gaps), hold them for a period of time and then decline in density as trees come to dominate a site (*e.g.* Putz *et al.*, 1984; Balée & Campbell 1990). This would mean that mainly gaps, especially large gaps, would be important for the establishment of lianas and that density of lianas should decrease with time after logging. Especially

Greenheart dominated mixed forest is suited to test this hypothesis, this main timber tree is growing in 'reefs', clustered cohorts, and gaps in logged over forest are mostly produced by cutting more than a single tree, resulting in quite a number of larger, multiple event, gaps (Ter Steege *et al.* 1996; Van der Hout 1996). Although the importance of large gaps for the establishment of species and individuals of lianas is true for all subplots together (table 3), and for the logged subplots, where the amount of large gap area leads to significant increase of species numbers and number of individuals, this relation is not significant for the unlogged plots. These results suggest that unlogged plots have a different liana vegetation composition, with liana species that prefer, or can endure, lower light levels, while in the logged sites species become dominant with preferences for higher light levels. This notion is enhanced by the observation that for the unlogged plots (table 3) a zone 5 meter out of gaps is important for an increased number of species and individuals, while in the logged plots a zone 5 meter into the gap is important. A young climber needs to be able to reach the forest canopy and a support structure is necessary. Appropriate trellises (supports) are most commonly found where there is regeneration of vegetation following disturbance and especially around the edges of gaps these supports are available in climbable sizes in the form of seedlings and saplings already present before the disturbance opened the canopy (after Balfour & Bond 1993). Species in the unlogged plots seem to favor trellises outside gaps, while species in logged plots seem to favor trellises in higher light conditions, inside the gap. It should be noted, however, that time of gapformation is not known for the unlogged plots. The gaps in unlogged plots could be much older than the ones appearing in the logged forest, and it could, therefore, be argued that the differences in preferences of species and individuals could represent forest recovery; where a gap closes due to lateral ingrowth of surrounding trees, and with liana establishment occurring mainly inside the gap. A recent research in French Guiana (Van der Meer & Bongers 1996b), however, showed that large gaps mainly fill through growth of gap floor regeneration, in which case no lateral shift of gap boundaries would occur.

Unlogged subplots, compared to logged subplots, show higher correlations with the higher illuminated zones of the fixed distance zonation (table 3), suggesting that in the unlogged plots light plays a more important role, being more limiting.

Incorporating the behavior of individual species (table 5) reveals that, over all subplots together, only 7 species show a significant increase in number of individuals within large gaps, while many correlate neutral, or even negatively, with gaps or gap size. The behavior of individual species (table 5, Fig. 3) shows that although lianas are mostly claimed to be pioneer species (*e.g.*, Putz, 1984; Peñalosa, 1984), some clearly decline in numbers or disappear completely as a result to logging, others proliferate and a third group reacts indifferently to logging. There are also no indications (table 1), that density of lianas decreases with time after logging, at least not on a time-scale of ten years after logging, and in fact we found a positive correlation between the time after logging and the number of individuals (table 2). This again, however, might be an effect from that fact that the oldest logged site had the highest species richness.

Skid trails

As stated before, on the level of one-ha plots the number of species after logging was positively correlated with the number of species present before logging, and negatively correlated with the amount of logging damage, expressed in number of harvested trees and skid trail area. Number of newly established species was positively correlated with overlapping gap and skid trail area. These results indicate that the amount of skid trail has a major impact on shifts in species composition. All initiatives to preserve species composition in relation to logging should try to minimize both number of felled trees and amount of skid trail.

Although on the level of total one-ha plots the correlations with skid trail area are negatively, on the level of subplots we find significantly positively correlations with number of species and individuals (table 3), mostly due to new, invading species. On skid trails we find a group of species with increased number of individuals specifically within this forest zone, together with species that correlate significantly with all damage classes, *e.g. Pinzona coriacea* and *Passiflora* spp. (table 5). Although the canopy over skid trails may stay closed, it is at least thinned out at lower levels, leading to higher light availability at the forest floor. Both the compacted and overturned soil forms a new establishment and germination site for species, due to e.g. changes in water retention capacity (Jetten 1994) and short-term changes in nutrient availability by decomposition of the overrun vegetation (Brouwer 1996). Species that correlate solely with skid trails are: *Tontelea coriacea, Machaerium madeirense,* and *Moutabea guianensis.* Others, like *Machaerium myrianthum* and *Strychnos bredemeyeri* correlate significantly both with overlapping skid trail and gaps and with gaps, suggesting that these species prefer even higher light levels. This notion is confirmed by the positive correlations with the higher light availability inner gap zones of the fixed distance zonation (table 5). *Curarea candicans* and *Mascagnia sepium* prefer the overlapping zone between skid trail and gap, while at the same time they do not show correlations with either gap or skid trail, although *Mascagnia sepium* shows a preference for the zone '5m in'. Perhaps these species prefer the higher light availability or trellis availability combined with higher water retention levels that can occur on skid trails.

Forest interior

Two years after logging (Fig. 2a) the largest increase in number of species is found outside gaps and mostly in an area more than five meters from a gap edge and even on sites with the lowest increased light levels. These sites in the forest also show quite an increase in number of individuals (Fig. 2b). This trend can be observed both for individuals with a height>2 m, as for smaller individuals (between 50 cm and 2 meters) (data shown for smaller size class). The larger individuals are mainly sprouts from fallen and damaged pre-logging lianas, whereas the smaller individuals are newly established. These results indicate that shortly after logging, especially the forest sites outside gaps, with (slight) increased light levels but no surface disruption, have favorable conditions for liana seedling establishment, both for species already present before the logging as for newly arriving species. One of the plots, logged2, however, shows a decline in number of individuals in the interior (more than 15 meters from gap edge), this is probably due to the fact that this plot has a rather high total gap area, 3606 m^2 (table 7), and has hardly any area left with a distance more than 15 meter from a gap edge (481 m^2). Those species that have a preference for low light intensities will decline in this plot. The only species that showed such a preference significantly for all subplots together is *Pleonotoma albiflora* (table 5), others such as *Dalechampia olympiana,* and *Forsteronia guyanensis,* show positive correlations with the lower light classes, but not significantly.

Logging in the Pibiri site was done in the first half of 1994, while skidding was separately undertaken in the second half of 1994 (Van der Hout 1996). New establishment of the vegetation after logging was delayed due to a very long dry season in 1994 through the beginning of 1995 (pers.obs). The patterns in liana vegetation visible two years after logging are merely a sign of the firsts invading species, together with the still present pre-logging vegetation. No severe competition or deaths due to the changed microclimate are expected to have taken place in this time frame. The pattern found shows that the first changes in liana diversity occur both outside gaps and inside gaps.

Seemingly in contrast, correlations between the amount of outer gap area and number of individuals or species are negative for Pibiri (table 4). Only species numbers in gaps show a significant positive correlation. This correlation, however, is calculated as a product moment over all 50 subplots, meaning that with increasing amount of outer gap area in a subplot there are less species and/or individuals, and with increasing amount of total gap area there are more species, compared to other forest zones. This means that in gaps there are more species per area, an indication that the density of liana species is higher within gaps, whereas density of species in outer gap zones is lower. When we compare the total amount of those two areas within the two one-ha plots (table 7), we see that for both plots the outer gap zone encompasses a much larger area than the total gap areas. Gaps do show an increase in species numbers (and individuals) (table 5) and over a smaller area this may lead to higher species densities.

In later phases of forest recovery (table 4), there are significant negative correlations between species numbers and numbers of individuals with the outer gap zone and the interior, suggesting that newly logged forest harbors a higher species diversity and abundance in those zones than older logged forests and that even in the 'undisturbed' forest areas, the patterns of diversity will change in time due to the logging.

Functional diversity

Four (4) different liana groups could be identified, each with a different response to logging:

1. Invading species, mainly appearing after logging and mostly establishing in large gaps and on skid trails.

This group of species could be considered as 'pioneer' or 'higher light preferentialists'. Most of these species have established themselves after the gap formation or proliferated after the logging and it can be expected that they germinated under high light levels. Examples: *Pinzona coriacea, Passiflora glandulosa.*

2A Indifferent species, with a preference for higher light levels, and mostly found around gap edges. Examples: *Lonchocarpus negrensis, Anomospermum grandifolium.*

2B Indifferent species, with a preference for lower light levels, mostly found at a distance of at least 10 meters from a gap edge. These species, although they probably first decline after the logging due to the increased light levels, are able to maintain comparable densities compared to the unlogged forest. Examples: *Connarus perrottetii, Forsteronia guyanensis.*

3 Exploitation sensitive species, regenerating in undisturbed forest patches and disappearing or declining in numbers after logging.
These species could be considered either as 'climax' species with a high preference for low light levels or as the group of species that have low densities and/or are poor competitors, in such a way that they either will be out competed by stronger competitors or have a low chance of being at a favorable spot where a gap is formed during logging. Most species that decline after the logging were found to be site specific and occurred in low densities in the unlogged sites. This finding would supports the latter notion. Two species, however, were abundant throughout all sites and could be considered 'true climax' species: *Pleonotoma albiflora, Memora moringifolia.*

We hypothesize that: 1) these groups can be considered as functional groups (guilds) which possess similar characteristics and adaptations to a changing environment; and 2) the responses of these functional groups to logging may be correlated to the responses of the total plant community. If this last hypothesis is valid, than the liana species composition can be used to assess the impact of logging damage on the biodiversity. Further research is needed to test this new hypothesis.

Acknowledgments

We would like to thank Demerara Timbers Ltd. for their hospitality and cooperation during our subsequent stays in the field. The work in the field could never have been successful without the help of several field assistants: G. Roberts, M. Nathan, M. Williams & M. Allicock are thanked for their assistance and company. B. Gravendeel is thanked for her contribution to the data of 'Pibiri' and help with the initial set-up, N. Raes and M. de Lind Van Wijngaarden for their contribution to the data of 'Reserve' and '2km'. The identification of the sterile reference material was especially vital for this research. Many people contributed to this laborious work and in particular J.C. Lindeman, B. Gravendeel, P.J.M. Maas, M.J. Jansen-Jacobs, S.A. Mori, M. Nee, B. Hammel and R. Liesner. For the help with identification of specific plant families we would like to thank G. Aymar (Dilleniaceae), A.R.A. Görts-van Rijn & A.M.W. Mennega (Hippocrateaceae), R. Barneby (Leguminosae and Menispermaceae), P. Hiepko (Menispermaceae), O. Poncy (Aristolochiaceae), P. Delgado (Rubiaceae) and C. Feuillet (Passifloraceae). The algorithms used for calculating the gap zones according to percentage of full sunlight were provided by A.J. Noest

(Bioinformatics Group, University of Utrecht), his help is greatly acknowledged. Furthermore, we would like to thank P. Hogeweg, M.J.A. Werger, S. de Jong, H. Ter Steege, R. Zagt and W. Dijkman for their help and stimulating discussions, and P. Hogeweg, H. Ter Steege and M.J.A. Werger for their comments on an earlier version of this manuscript. This study was carried out in the framework of the Tropenbos Guyana Programme and funded by the Tropenbos Foundation, Wageningen, the Netherlands, Utrecht University, and the EU.

Appendix 1:

Equations used for calculation coordinates within the plots and the eight gap edge points

AB:	$\alpha_c = \alpha_m$	$X = Y * \tan \alpha_c$
	$\beta_c = 180° - \beta_m$	$Y = AB * \tan \beta_c / (\tan \alpha_c + \tan \beta_c)$
AD:	$\alpha_c = \alpha_m - 90°$	$X = AD * \tan \beta_c / (\tan \alpha_c + \tan \beta_c)$
	$\beta_c = \beta_m - 270°$	$Y = X * \tan \alpha_c$
BC:	$\alpha_c = 90° - \alpha_m$	$X = BC * \tan \beta_c / (\tan \alpha_c + \tan \beta_c)$
	$\beta_c = 270° - \beta_m$	$Y = BC - X * \tan \alpha_c$
CD:	$\alpha_c = 360° - \alpha_m$	$X = CD - Y * \tan \alpha_c$
	$\beta_c = \beta_m - 180°$	$Y = CD * \tan \beta_c / (\tan \alpha_c + \tan \beta_c)$

$\alpha_c = \alpha$ corrected , $\alpha_m = \alpha$ measured - 13°

$\beta_c = \beta$ corrected, $\beta_m = \beta$ measured - 13°

Appendix 5.2:
Liana species found at four sites near Mabura Hill, Guyana, with their authorities.

APOCYNACEAE
Forsteronia acouci (Aubl.) A. DC.
Forsteronia gracilis (Benth.) Muell. Arg.
Forsteronia guyanensis Muell. Arg.
Forsteronia schomburgkii A. DC. cf.
Odontadenia amazonica (Stadelm.) Malme
Odontodenia geminata (Roem. & Schult.) Muell. Arg. cf.
Odontadenia puncticulosa (A. Rich.) Pulle

ARACEAE
Heteropsis flexuosa (Kunth) Bunting
Philodendron rudgeanum Schott
Rhodospatha venosa Gleason

ARISTOLOCHIACEAE
Aristolochia consimilis Mast.
Aristolochia daemoninoxi Mast.
Aristolochia rugosa Lam.

BIGNONIACEAE
Anemopaegma oligoneuron (Sprague & Sandw.) A. Gentry
Anemopaegma parkeri Sprague
Arrabidaea egensis Bureau ex K.Schum.
Arrabidaea fanshawei Sandw.
Arrabidaea mollis (Vahl) Bureau ex K.Schum.
Clytostoma binatum (Thunb.) Sandw.
Clytostoma sciuripabulum Bureau & K. Schum.
Cydista aequinoctialis (L.) Miers
Distictella elongata (Vahl) Urb.
Memora bracteosa (DC.) Bureau & K. Schum. cf.
Memora flavida (DC.) Bureau & K. Schum.
Memora moringifolia (Miq.) Pulle
Memora racemosa A. Gentry
Memora schomburgkii (DC.) Miers
Mussatia prieuriei (DC.) Bureau ex K. Schum.
Pleonotoma albiflora (Salzm. ex DC.) A. Gentry
Roentgenia sordida (Bureau & K. Schum.) Sprague & Sandw.
Schlegelia violacea (Aubl.) Griseb.

CECROPIACEAE
Coussapoa microcephala Trécul

COMPOSITAE
Mikania gleasonii B.L. Rob.
Mikania sp.
CONNARACEAE
Cnestidium guianense (Schellenb.) Schellenb.
Connarus coriaceus Schellenb.
Connarus erianthus Benth. ex Baker var. stipitatus Forero
Connarus megacarpus S.F. Blake
Connarus perrottetii (DC.) Planch. var. rufus Forero
Connarus punctatus Planch
Pseudoconnarus macrophyllus (Poepp. & Endl.) Radlk.
Rourea induta Planch. var. induta
Rourea ligulata Baker cf.
Rourea pubescens (DC.) Radlk. var. spadicea (Radlk.) Forero

CONVOLVULACEAE
Dicranostyles guianensis Mennega
Lysiostyles scandens Benth.
Maripa glabra Choisy cf.
Maripa scandens Aubl.

CUCURBITACEAE
Cayaponia ophthalmica R.E. Schultes
Cayaponia selysioides C. Jeffrey
Gurania bignoniacea (Poepp. & Endl.) C. Jeffrey
Gurania spinulosa (Poepp. & Endl.) Cogn.
Helmontia leptantha (Schltd.) Cogn.

CYCLANTHACEAE
Evodianthus funifer (Poit.) Lindm. spp. trailanus (Drude) Harling

DICHAPETALACEAE
Dichapetalum pedunculatum (DC.) Baill.
Dichapetalum rugosum (Vahl) Prance

DILLENIACEAE
Doliocarpus brevipedicellatus Garcke ssp. brevipedicellatus
Doliocarpus guianensis (Aubl.) Gilg.
Doliocarpus macrocarpus Mart. ex Eichler
Doliocarpus major J.F. Gmel. ssp. major

Doliocarpus paraensis Sleumer
Pinzona coriacea Mart. & Zucc.
Tetracera asperula Miq.
Tetracera volubilis L. ssp. volubilis

DIOSCOREACEAE
Dioscorea dodecaneura Vell.
Dioscorea megacarpa Gleason cf.

EUPHORBIACEAE
Croton sp.
Dalechampia olympiana Kuhlm. & Rodr.
Dalechampia parvibracteolata Lanj. cf.

GNETACEAE
Gnetum nodiflorum Brogn.
Gnetum urens (Aubl.) Blume

GUTTIFERAE
Clusia grandiflora Splitg.
Clusia myriandra (Benth.) Planch. & Triana
Clusia pana-panari (Aubl.) Choisy

HIPPOCRATEACEAE
Cheiloclinium cognatum (Miers) A.C. Sm.
Cheiloclinium hippocrateoides (Peyr.) A.C. Sm.
Peritassa laevigata (Hoffmanns.) A.C. Sm. cf.
Prionostemma aspera (Lam.) Miers
Salacia multiflora (Lam.) DC. ssp. multiflora
Tontelea attenuata Miers
Tontelea coriacea A.C. Sm. cf.
sp1
sp2

ICACINACEAE
Leretia cordata Vell.

LEGUMINOSAE/CAESALPINIODEAE
Bauhinia guianensis Aubl. var. guianensis
Bauhinia scala-simiae Sandw.
Bauhinia sp.
Bauhinia sp2.
Bauhinia surinamensis Amsh. cf.

LEGUMINOSAE/PAPILINOIDEAE
Clitoria sagotii Fantz. var. sagotii
Clitoria sagotii Fantz var. canaliculata Fantz
Dioclea elliptica R.H. Maxwell
Dioclea scabra (Rich.) R.H. Maxwell
Lonchocarpus negrensis Benth.
Machaerium madeirense Pittier
Machaerium multisii Killip ex Rudd
Machaerium myrianthum Spruce ex Benth.
Machaerium oblongifolium Vogel cf.
Machaerium quinata (Aubl.) Sandw. var. quinata
Machaerium sp

LOGANIACEAE
Strychnos bredemeyeri (Schult.) Sprague & Sandw.
Strychnos diaboli Sandw. cf.
Strychnos erichsonii M.R. Schomb.
Strychnos hirsuta Spruce ex Benth.
Strychnos melinoniana Baill.
Strychnos subcordata Spruce

LYGODIACEAE
Lygodium volubile Sw.

MALPIGHIACEAE
Banisteriopsis martiniana (A. Juss.) Cuatrec. var. martiniana
Heteropterys cristata Benth. cf.
Heteropterys multiflora (DC.) Hochr.
Hiraea adenophora Sandw. cf.
Hiraea affinis Miq.
Hiraea fagifolia (DC.) A. Juss.
Mascagnia sepium (A. Juss.) Griseb.
Mezia includens (Benth.) Cuatrec.
Stigmaphyllon sinuatum (DC.) A. Juss
Tetrapetrys sp.
Tetrapterys calophylla A. Juss.
Tetrapterys crispa A. Juss.
sp1

MARCGRAVIACEAE
Marcgravia parviflora Rich. ex Wittm.
Norantea guianensis Aubl.
Souroubea guianensis Aubl. ssp. guianensis

MENISPERMACEAE
Abuta barbata Miers
Abuta bullata Moldenke
Abuta imene Eichler
Abuta rufescens Aubl.
Abuta sandwithiana Krukoff & Barneby
Anomospermum grandifolium Eichler

Curarea candicans (Rich.) Barneby & Krukoff
Telitoxicum krukovii Moldenke
Telitoxicum minutiflorum (Diels) Moldenke

MORACEAE
Ficus guianensis Desv. cf.

ORCHIDACEAE
Vanilla cristato-callosa Hoehne cf.
Vanilla sp.

PALMAE
Desmoncus macroacanthos Mart.

PASSIFLORACEAE
Dilkea wallisii Masters
Passiflora auriculata Kunth
Passiflora cirrhiflora A. Juss.
Passiflora coccinea Aubl.
Passiflora fuchsiiflora Hemsl.
Passiflora garckei Masters
Passiflora glandulosa Cav.
Passiflora misera Kunth
Passiflora sp1 (prob. spec. nov.)
Passiflora vespertilio L.

PIPERACEAE
Piper hostmannianum (Miq.) C. DC.

POLYGALACEAE
Moutabea guianensis Aubl.
Securidaca spinifex Sandw.

POLYGONACEAE
Coccoloba lucidula Benth. cf
Coccoloba marginata Benth. cf.
Coccoloba parimensis Benth.
Coccoloba sp.

RUBIACEAE
Malanea macrophylla Bartl. ex Griseb. var. macrophylla
Malanea sp.
Randia asperifolia (Sandw.) Sandw.
Sabicea surinamensis Bremek. cf.
Uncaria guianensis (Aubl.) J.F. Gmel.

SAPINDACEAE
Paullinia capreolata (Aubl.) Radlk.
Paullinia ingaefolia Rich.
Paullinia pachycarpa Radlk.
Paullinia sp1
Serjania paucidentata DC.

SMILACACEAE
Smilax cumanensis Willd.
Smilax poeppigii Kunth
Smilax santaremensis DC.
Smilax syphilitica Willd.

SOLANACEAE
Markea coccinea Rich.
Markea sessiliflora Ducke
Solanum adhaerens Roem. & Schult.

VERBENACEAE
Petrea macrostachya Benth.
Petrea volubilis L.

Family unknown
undet1- undet 4

CHAPTER 6

Patterns of species diversity in the neotropical lowland rain forest of Guyana

Many different theories exist that try to explain the present day levels of species richness in tropical rain forests. One of the main problems is that the spatial and temporal scales at which these hypotheses are valid is seldomly identified. Testing of these hypotheses, however, is only possible when the spatial and temporal scales are explicitly defined. The processes that lead to present-day levels of (plant) species richness are considered to be a combination of both historical (origin of diversity) and present-day ecological processes (maintenance of diversity) (Fig 1.1).

Historical models include two biogeographical theories concerning differences in species richness and levels of endemism in the tropics: the Pleistocene refuge hypothesis (Haffer 1974, 1982, 1997; Prance 1982, 1985) and the Pleistocene cooling hypothesis (Colinveaux 1993) (level A, Fig. 1.1). These hypotheses mention historical disturbances as important for past speciation and present-day patterns of species distribution. Another theory that can be considered in historical context, the stability - time hypothesis (after Sanders 1969), mentions the non-disturbance status of tropical forest as an important factor for determining present-day species richness levels (level B, Fig. 1.1).

For the present-day ecological processes, operating on smaller spatial and temporal scales than the previously mentioned hypotheses, both equilibrium and non-equilibrium theories have been proposed. Equilibrium theories view tropical community as a balanced community of many species whose coexistence is explained based on Lotka-Volterra competition and predation equations for species that require the same resources (*e.g.*, MacArthur & Wilson 1967; Janzen 1970; Whittaker 1967). Tropical species then are considered as highly specialized, *i.e.*, having different adaptations to a specific optimum combination of conditions of the environment (both abiotic and biotic) (level B, C & D, Fig. 1.1). Each of the species is considered as competitively superior over all others within a specific combination of environmental conditions. This differential specialization would lead to a deterministic outcome of species interactions and successional processes. Without disturbances or within a homogeneous environment a rather small set of species would be able to outcompete all other species. A high level of diversity is explained by heterogeneity of the environment, or regular disturbances. The gap partitioning theory, the intermediate disturbance hypothesis, and the compensatory mortality theory are examples of such equilibrium theories (*e.g.*, Huston 1994; Brokaw 1985; Hartshorn 1978; Whitmore 1984; Organs 1982; Zag & Werger in press.). In contrast, non-equilibrium theories do not emphasize such differential specialization in species. For instance, Hubbell (1979), Benzing (1991) and Hubbell & Foster (1986), proposed that the existence of a high number of plant species in any tropical rain forest patch is promoted by the ecological and competitive similarity of the species, which reduces the chances of competitive exclusion. This type of hypothesis is also referred to as 'random walk hypotheses'.

In this chapter, some of the results as presented in this thesis, will be used in order to evaluate the applicability of the mentioned theories for tropical rain forest in Guyana. It should be noted, however, that a conclusive evaluation might not be possiblė. Even if deterministic processes play an important role in determining the species composition at a given site, chance processes such as gap formation and presence of advanced regeneration at the sites of gap formation (Hubbell & Foster 1986; Brokaw 1986; Van der Meer & Bongers 1996a; Zagt & Werger in press), may hamper recognition of such deterministic patterns.

The Pleistocene refuge hypothesis and the Pleistocene cooling hypothesis

According to the Refuge hypothesis, centers of endemism reflect forest refuges in the peripheral regions of Amazonia during dry periods of the Tertiary and Quaternary (Haffer 1974, 1982, 1997; Prance 1982, 1985, 1989). Centers of endemism, and present-day species richness, would then correspond to ecologically stable areas that retained their rain forest cover throughout the Pleistocene. In these environments that stayed wet in surrounding, drier vegetation, rain forest plant and animals formed isolated populations, which gave rise to rapid speciation processes (vicariance). The 'refuges' functioned as source areas from where plant and animal populations spread when the climate changed. In contrast, in the Pleistocene cooling hypothesis (Colinveaux 1993), the peripherical regions of Amazonia are considered ecologically instable. In this theory, the most important environmental factor leading to vicariance of species is cooling rather than aridity. The peripheral regions of Amazonia are seen as dynamic borderlands between the mountainous areas and lowlands, and intense competitive species interactions are assumed to have favoured species isolation.

Ever since the formulation of these hypotheses, they have been much disputed. The Refuge hypothesis has been criticized on several grounds (for an overview see Tuomisto & Ruokolainen 1997), with as main arguments that the centers of endemism are not the same for different species groups, although some show some overlap (*e.g.*, for trees, butterflies, and birds) (Beven *et al.* 1984); that they may be artifacts of collecting intensity (Nelson *et al.* 1990); and that most species of lowland rain forest predate the Pleistocene (Fjeldså & Lovett 1997). Haffer (1997), however, cites new evidence supporting the predictions of the refuge theory. In the same paper, Haffer states that the Pleistocene cooling theory cannot explain speciation in the Tertiary, when primarily dry/humid cycles and basically no cold/warm cycles occurred. Furthermore, according to some authors (*e.g.*, Huston 1994; Tuomisto & Ruokolainen 1997) it is also possible to explain the patterns that would result from the refuge theory and the Pleistocene cooling hypothesis by present-day environmental gradients. They postulate that pursuing explanations of distribution and speciation patterns in historical context, still assumes that tropical forests in Amazonia, and those in its peripherical regions, are ecologically uniform. Only if ecological environments in tropical forests could be considered very homogeneous, would it be valid to assume them insignificant in determining present-day species distribution and past speciation patterns. Recent studies, however, showed ecological heterogeneity and floristic differences among tropical rain forest sites that were previously thought similar (Tuomisto & Ruokolainen 1997). The only way to assess the relative importance of present-day ecological and historical factors in explaining a particular pattern of species occurrences is to find out how much of the observed patterns remain unexplained after the effects of current ecological factors have been taken into account (Borcard *et al.* 1992).

The study area, Mabura Hill, lies within one of the postulated Pleistocene forest refuges for trees, as proposed by Prance (1982, 1985). To validate this hypothesis paleobotanical data are needed, and no data originating from the present study can be used. It is hoped, however, that this study contributes to explanations of present-day species distribution patterns within Greenheart-dominated forest.

The stability - time hypothesis

Under this hypothesis the long history of tropical forests and the relative constant humidity and climate or homogeneous environment, induced either extreme specialization (*e.g.*, Ashton 1969, 1993) or ecological and competitive similarity of species (*e.g.*, Hubbell 1979; Benzing 1981; Hubbell & Foster 1986), which resulted in a high diversity in tropical regions compared to ecosystems in temperate regions. Differences in viewpoints on whether competition and succession are deterministic processes or governed by chance factors, and on how species partition specific ecological niches (theoretical possible distribution of a species along a gradient), underlay these different viewpoints (for an overview see Huston 1994; Zagt & Werger in press)

At the moment that the stability - time hypothesis was originally formulated, the tropical rain forest was still regarded as a stable and unchanging environment. As stated before, however, the tropical rain forest is recently found much more heterogeneous than previously assumed (Tuomisto & Ruokolainen 1997). According to Huston (1994) the stability - time model might explain some diversity patterns over evolutionary time scales in situations where the extinction rate is very low and at very short (dispersal) time scales.

As regards the present study, the stability-time hypothesis might be an explanation for the diversity levels of tropical rain forest of the Guianas. The Guianas, in contrast to Amazonia and the Caribbean, did not experience frequent large-scale disturbances (Hammond & Brown 1995; Berry *et al.* 1995). Sheltered by the Caribbean islands and situated on the Guayana Shield, the Guianas experience no hurricane events or other large-scale windthrows, no volcanoes, or large scale river dynamics. In this thesis (chapter 2), it is argued that the origin of the most important families in the Mabura Hill area, the large proportion of Caesalpinoideae in the Guayana Shield and Northern Brazil sites, and the high percentage of such families as Connaraceae, Menispermaceae, Nyctaginaceae, Annonaceae, and Apocynaceae support the hypothesis that the flora of the Guianas shows a strong link with the forest flora of Western Africa, stronger than other parts of South America. It was argued that the visibility of this link might be explained by the low disturbance status of the area. Furthermore, Hammond & Brown (1995) showed that the mean seed weight of 187 native woody species of the Mabura Hill area, both trees and lianas, was significantly larger than that found in two other sites in South America, Manu and Barro Colorado Island (table 6.1). These authors postulated that the percentage of large-seeded species in a flora is correlated with the disturbance status of an area. Additionally, the possible correlation between the presence of large-seeded species and disturbance was supported by comparing three Guianan sites with different levels of historical disturbance. The most disturbed site, Saül, had the lowest proportion of large-seeded tree families, the intermediate disturbed site, Nouragues, an intermediate proportion of large-seeded tree families, while Mabura Hill had the highest proportion. Furthermore, although the stability-time hypothesis concerns regional levels of diversity, at the local level it would be expected that Greenheart forest belongs to a mediate species-rich to species-poor tropical rain forest type (chapter 4) if the area hardly experienced disturbance (see also below, intermediate disturbance hypothesis).

Table 6.1
Difference in mean seed weight for Mabura Hill, Manu and Barro Colorado Island. Data from Hammond & Brown (1995).

	Mabura		Manu		BCI	
	mean	stdev	mean	stdev	mean	stdev
FRESH MASS						
All species	7.52	1.44	1.68	0.44	0.92	0.38
Leguminosae	13.42	4.64				
Caesalpinoideae	27.81	12.66				

Particularly if the estimate of Zagt (1997) is valid, that the age at which a Greenheart tree becomes reproductive is 200 years.

In the previous section only circumstantial evidence is presented, concerning stability of the Mabura Hill area at a large time scale. This does not give an indication whether this stable environment led to extreme specialization (equilibrium model) or ecological and competitive similarity of species (non-equilibrium model). However, under the assumption that a stable environment over a long time leads to extreme specialization of species, it can be hypothesized that species in Greenheart forest specialized for this vegetation type and that the different plots within Greenheart forest would have approximately the same species composition. In chapter 4 (table 4.6), similarity of all 24 one-ha plots using the Jaccard Index, showed that there are considerable differences in species composition between the plots, and that in general only 50% of the species overlap between plots. Furthermore, the similarity between sites, plots (table 4.6), and within plots (table 4.7), showed to be distance dependent. These results do not support this extreme specialization hypothesis .

In the future, the permanent sampling plots in the Pibiri area may facilitate testing if the stability-time hypothesis can be applied at shorter time frames. Under the equilibrium assumption that a stable environment would lead to extreme specialization it can be hypothesized that the unlogged referential plots in Pibiri will have approximately the same species composition between the first and second enumeration. Alternatively, under the non-equilibrium assumption that a stable environment would lead to ecological or competitive similarity of species it can be hypothesized that between the first and second enumeration of the unlogged referential plots in Pibiri, the change in species composition is stochastic, in particular of those plants that have established since the first enumeration. However, changes on temporal and spatial scales are sometimes related differences, *e.g.*, the life-spans of individuals and the rates of dispersal. Observed changes in species composition may therefore seem stochastic, while in reality they are largely the result of deterministic processes such as dispersal and survival. Thus, observed stochastic patterns do not necessarily support the assumption that species are similar in ecological or competitive abilities.

These hypotheses can only be evaluated in a later phase of the Tropenbos Guyana Programme, after a complete second enumeration of the Pibiri plots.

The gap partitioning hypothesis

Gap partitioning occurs when species are niche-differentiated and, through differences in growth, mortality, and/or reproduction requirements, regenerate optimally in gaps of different sizes or in different environments within a gap, effectively partitioning them amongst each other (Denslow 1980; Brokaw 1985a, 1985b, 1987; Hartshorn 1978, 1980; Whitmore 1984, 1978; Orians 1982).

Under the assumption that species are niche-differentiated it can be hypothesized that within the plots there may be a significant different species composition between gaps of different sizes. As the moment of gap creation is important for establishment probabilities of species, it is expected that particularly in the logged plots (Waraputa, Reserve and 2km) this hypothesis can be tested. In those plots colonization of gaps is expected to have started since the time of clearing, regardless of size. In chapter 4 (table 4.9), it was documented that the species composition between large (≥300 m^2) and small (<300 m^2) gaps of the same exploited one-ha plot was significantly different for four out of six logged plots. These results would suggest that gap partitioning between small and large gaps might play a role. However, in chapter 4 it was also shown that significant differences in species composition existed between gaps of the same size. Separate clustering of damage zones even showed that the sites always clustered together, both the 'undisturbed' forest patches and the gaps. This indicates that the species composition that fills a gap is mainly determined locally.

It has been documented that especially the increased light levels in and around gaps are most important in explaining variance in seedling establishment and performance (*e.g.*, Connell 1978; Brokaw 1985; Popma & Bongers 1988; Clark & Clark 1990; Welden *et al.* 1991). Although in this study light was not directly measured, the orientation and size of gaps were used to calculate a hypothetical year average light gradient over the plots. These light gradients were computed under the assumption that light levels around gaps were influenced by the size and orientation of the gap towards the solar track, combined with the height of the surrounding canopy. Also taken into account was the distance from the gap edge and the density of surrounding vegetation as determinants of the attenuation of light into the surrounding vegetation. For all points in and around gaps the time of exposure to direct light through the gap as a percentage of full sunlight were calculated (for equations used, see appendix 6.1). This hypothetical light gradient would only be valid for a period of time shortly after gap formation. In a later stage of forest regeneration, with new vegetation formed within the gaps, the light gradient will have changed considerably. Consequently, the calculated light gradient could only be applied to those individuals which might be assumed to have established shortly after logging. Liana individuals in the size class with a height of more than 2 meters were expected to qualify. This assumption was supported by the fact that newly established species after logging were mainly found in this size class (table 4.8, 5.1). In chapter 5, three main groups of liana species are described, each with its own reaction to logging: a group of species that proliferated after logging (group1), a group of species that reacted indifferent to logging (group 2), and a group of species that disappeared as a result of logging (group 3). When gap partioning would result from differences in light climate, it may be expected that the three groups of liana species would show different preferences for light environments. The first group (proliferating species) is expected to show a preference for high light environments, the second group (indifferent species) is expected to show no particular preferences, and the third group (exploitation sensitive species) is expected to show a preference for low light environments (or older successional stages). To test this, for each group two example species were used; *i.e.*, the two most abundant species of each group. For each species, the occurrences of all individuals with a height of more than 2 meters within the computed light percentage classes were used, over 9 one-ha plots (exploited forest sites). The increased light percentages were pooled into logarithmic classes, due to the fact that in the lower light environments a small increases in light availability is much more important for plant performance than in the higher light environments (Boot 1994; Chazdon *et al.* 1996). For more details on the method used, see chapter 4. The results are presented in Fig. 6.1. The species that represented the first group, *Pinzona coriacea* and *Passiflora* spp., showed a clear preference for higher light levels. In the second group, both *Connarus perrottetii* and *Anomospermum grandifolium,* showed no clear preference for specific light levels. *Connarus perrottetii,* however, occurred more often in lower light levels. This agrees with the results obtained from the fixed distance zonation (table 5.5) that this species had a slight significant correlation with the zones outside a gap, which resulted in the recognition of a group 2B (indifferent species with a preference for lower light levels). The observation that *Anomospermum grandifolium* occurred more in higher light environments also agrees with the results obtained ,in table 5.5, and the recognition of group 2A (indifferent species with a preference for higher light levels). Thus far, the results support the gap partitioning model. For the last group (exploitation sensitive species) the results were more ambivalent. It was clear, however, why those two species were labeled as negatively correlated with logging: both had a much higher number of individuals in unlogged plots compared to logged plots, notwithstanding that the latter had twice as much plots. One species, *Pleonotoma albiflora,* did show a preference for low light environments, although this observation was based on 7 individuals only. The other species, *Memora moringifolia,* showed a preference for higher light environments, both for unlogged and logged plots. It seems that the decrease of this species after logging cannot be attributed to the higher light levels after logging. An explanation for the decrease of *M. moringifolia* could possibly be found in the evaluation of where in the plots this species was found most often. In Fig. 6.2. the occurrence of this species over different damage zones is given. *M. moringifolia* seemed to prefer the gap edge zone, and secondly zones outside the gap (but with high light availability),

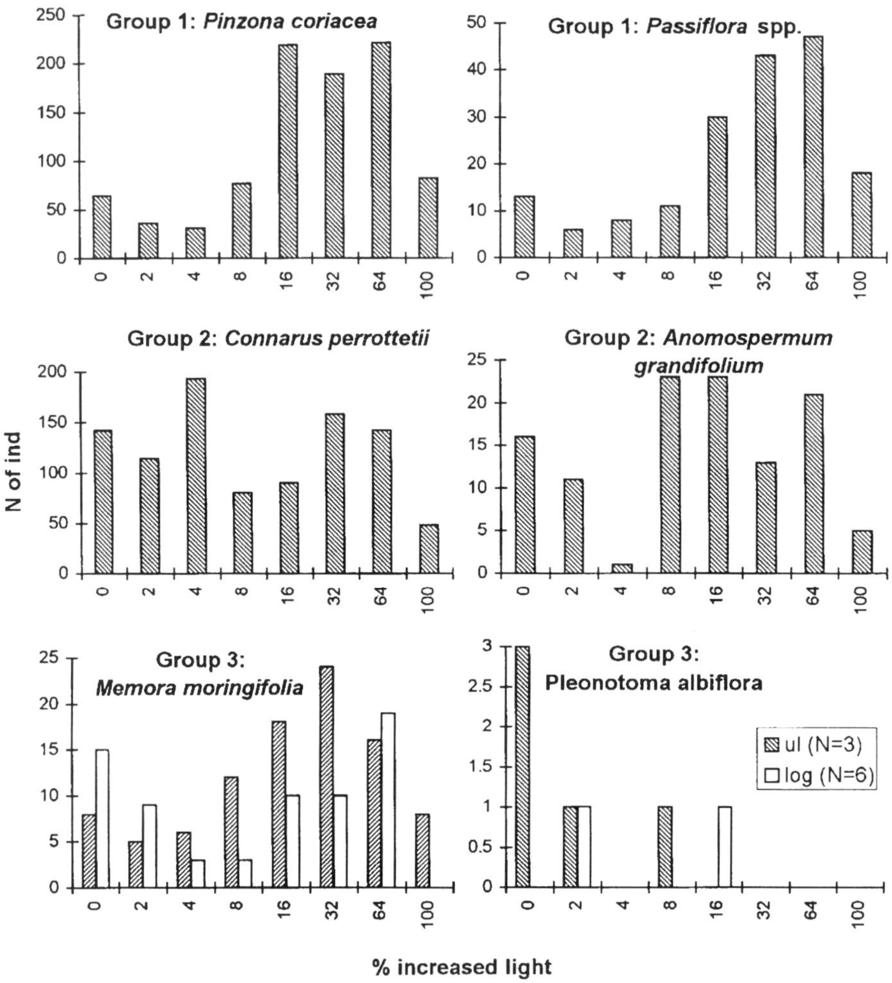

Figure 6.1
Distribution of individuals of six liana species over increased light levels in Waraputa, Reserve and 2km, all older logged areas in the vicinity of Mabura Hill, Guyana. Increased light percentages are pooled into logarithmic classes. Group 1: Proliferating species after logging; group 2: indifferent species; group 3: exploitation sensitive species.

so possibly those zones that showed higher light availability but no surface disruption. In logged plots there were far less zones with no surface disruption compared to the unlogged plots. This again, would indicate a clear preference of a species for a specific habitat in the forest. Another factor that could play a role in the decrease of *M. moringifolia* after logging might be chance. It is not known whether or not this species occurred in the plots before logging, and it might be possible that *M. moringifolia* did show an increase in number of individuals compared with the

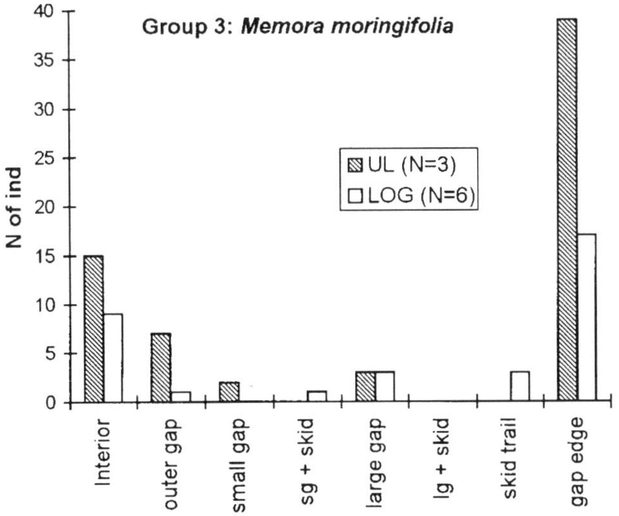

Figure 6.2
Distribution of individuals over different damage zones for *Memora moringifolia*, an exploitation sensitive liana species.

unlogged status, but that this is not noted due to high numbers of individuals of this species in the unlogged plots. However, only a small part of all liana species showed clear correlations (chapter 5). Also within the other growth-forms only a small percentage of the species showed clear preferences for a specific habitat (table 4.13). Most species occurred only locally, or were only locally abundant, and our results indicate that clear deterministic preferences of species are more an exception than a rule. In conclusion, the gap partitioning hypothesis seems to be supported by our findings only for some liana species.

The intermediate disturbance hypothesis

For the intermediate disturbance hypothesis (Connell 1978; Huston 1979) disturbance (and as a consequence, mortality) promotes the coexistence of species through reduction of competition by dominants. Given a range of disturbances of different intensities, the highest diversity will be expected at intermediate levels of disturbance. If the disturbance is too mild or occurs too rarely, patches will approach equilibrium and be dominated by a few species that are able to outcompete others. If the disturbance is too harsh or occurs too often, only a few species that are resistant to the frequent disruption will persist. Plant species adapted to high levels of disturbance are at a competitive disadvantage under low disturbance levels, because species adapted to high levels of disturbance have none of the constraints associated with survival in undisturbed circumstances (Huston 1994).

Under the assumption of this hypothesis, species diversity is expected to increase with increasing logging intensity until a certain level ('intermediate level') after which species diversity decreases with increasing logging intensity.

Evaluation of this hypothesis is mainly possible after a second enumeration of the Pibiri plots. The logging in the reduced-impact logging area has been conducted with three different logging intensities (Van der Hout 1996), which enables the evaluation of the effects of logging intensity. Is should be noted, however, that it is not known what is 'low', 'intermediate' or 'high' intensity in relation to logging in Greenheart forest. It might be possible that all logging (also with conventional techniques) is 'intermediate'. It also might be possible that these levels are different for different growth-forms.

At this moment, the preliminary observation can be made that species richness of lianas of two Pibiri plots (chapter 5), was higher after logging. The time scale at which the observations for the two Pibiri plots were made is too short for strongly competitive species to outcompete others, which is according to the intermediate disturbance model an important process that may lead to a decrease in diversity. The increase in species richness and abundance after logging was also found in the exploited forest sites for all growth forms (chapter 4 & 5). However, these results may be misleading due to the fact that the levels of diversity before logging are not exactly known, and that these sites may have been repeatedly disturbed. In conclusion, at this moment no definite answer can be drawn concerning the merit of the intermediate disturbance hypothesis.

The compensatory mortality theory

Comparable with the last hypothesis, compensatory mortality is considered as a disturbance which will set back the community from a possible equilibrium and will prevent dominant species to outcompete others (Connell 1978; Janzen 1970). Compensatory mortality is viewed in the form of frequency, density, or distance-dependent mortality (*e.g.* through herbivory or parasites). For instance, Janzen (1970) suggested that where seeds occur clumped, they are most heavily eaten. Recent theoretical investigations using a modified version of the Shmida & Ellner model (1984) have suggested that spatial density and frequency dependence need not be strong to have an important effect on the maintenance of tropical tree diversity (*e.g.* Hubbell & Foster 1992; Condit *et al.* 1994).

To test whether frequency, density or distance-dependent mortality operates within Greenheart forest, information on differences in growth, mortality, and predation of common and rare species is needed. Those data are at present not available.

The 'random-walk' hypothesis

The 'random-walk' hypothesis (Hubbell 1979; Hubbell & Foster 1986; Benzing 1991) is an example of a non-equilibrium model in which the probability of competitive exclusion is reduced by ecological and competitive similarity of species, thus resulting in a high species richness. Following this hypothesis, the species composition of any given site, that can be considered homogeneous, is a randomly drawn subset from the pool of species that is ecologically suitable to occupy the site. Turnover of species between plots of the same homogeneous site are expected to be mainly spatially controlled. As mentioned above, the finding of spatially controlled changes in species composition, however, does not necessarily support the random walk hypothesis. Patchiness can be the result of deterministic processes, such as life span of individuals and dispersal opportunities. A crucial aspect in this hypothesis is this term 'homogeneous'; which level of variation is still allowed for this hypothesis to be applicable? Another crucial aspect is that this hypothesis assumes that the process of competitive exclusion occurs very slow, so that competitive interactions between individuals are not strong, or equally strong.

Three levels of information are available to evaluate this hypothesis at different spatial scales; at the level of ground-rooted species occurring in different patches of Greenheart forest (between plots), for the same group of ground-rooted species within plots, and at the level of different epiphytic species occurring in different height zones in trees and between different forest types. On the level of the total flora of the Mabura Hill area it should be noted that the tropical rain forest of Guyana was found to be dominated by a few or one species over large patches, often associated with particular soil types or hydrological conditions (Davis & Richards 1932, 1934; Fanshawe 1952; Ter Steege *et al.* 1993). This indicates that the presence of different forest types in the surroundings of Mabura Hill is more deterministic than stochastic, or that the environment at this level of observation can not be considered as homogenous.

We start at the largest spatial level, the level of different epiphytic species occurring in different height zones in trees and compare different forest types. Vertical distribution of epiphytes is mainly determined by light, atmospheric humidity, and character of the substratum (*e.g.* Johansson 1974; Bennett 1986; Wolf 1994; Benzing 1989, 1990, 1995). It can be argued that most epiphytes require conditions provided by many kinds of trees and that when the requirements of light, moisture, and substratum availability are fulfilled, trees from different forest types can be considered as offering a homogeneous environment. Under the assumption of the random walk hypothesis, it is expected that only a small part of the variation in epiphytic species composition between forest types can be explained, and mainly through geographical variation. In chapter 3, however, it was shown that shared number of species and Jaccard's Index of similarity between three forest types in the Mabura Hill area (table 3.2), and one forest type in French Guiana, was always the highest with the French Guianan forest type. Furthermore, differences in epiphytic species richness between forest types seemed to be strongly correlated with humidity conditions, resulting in higher epiphytic species richness in more humid conditions. Within forest types, humidity conditions seem to be correlated with the vertical distribution patterns of epiphytic species on their hosts. These results indicate that epiphytic species did show

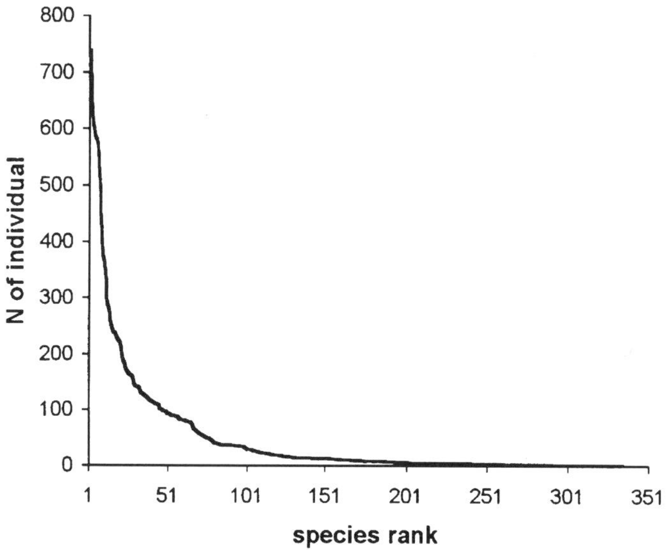

Figure 6.3
Observed distribution of individuals over species (all growth-forms) for 15 unexploited one-ha plots of Greenheart dominated forest near Mabura Hill, Guyana.

different adaptations to, or prefer a specific optimum combination of, conditions of the environment, and are thus determined by that environment. However, this result applies to the range over a tree in which a specific epiphytic species was found. According to the results of chapter 3, these ranges can be divided in four different habitats (appendix 3.2, Fig. 3.4), and only within those habitats one can really speak about a homogeneous environment. It may still be possible that the epiphytic species occurring in one of those four habitats are a random subset of all species adapted to grow within that habitat. The exact delineation of the four habitats is different within each forest type, depending on airhumidity and insolation (chapter 3). In each forest type, however, the first inner canopy zone (zone 4, Fig. 3.1) belongs to habitat C, the inner canopy. The species pool occurring in this zone in the richest forest type may be considered as the largest representative of the group of epiphytic species suited to occur in this habitat. Under the random walk hypothesis the species occurring in this zone in the other forest types are expected to be -to a large extent - a subset of that pool of species. Although all forest types showed the highest similarities with the richest forest type (*MIX*)(table 3.2), agglomerative clustering showed that the inner canopy zones of all forest types clustered in three different groups (Fig 4.5). A group of inner canopy zones of the two wettest forest types, Mora forest (*MOR CF*) and mixed forest (*MIX*); a group of inner canopy zones of all forest types together; and a group of inner canopy zones of only mixed forest (*MIX*). These results indicate that also within the inner canopy zone of different forest types the epiphytic environment cannot be considered as homogeneous and we need to make a distinction between an inner canopy zone in wetter and drier forest types. Even the inner canopy zones of one forest type (*MIX*) were found in three different clusters (Fig. 3.4). Thus, even within one forest type the inner canopy zone cannot be considered completely homogeneous, and the results suggest that the species composition is highly determined by moisture availability. These results do not support the random walk hypothesis, and would suggest that although random processes can play a considerable role, the epiphytic species composition is largely determined by local deterministic processes.

To evaluate the random walk hypothesis on the level of ground-rooted species, *i.e.*, for those species that are ecologically suitable to grow within Greenheart forest on Brown Sand, plots from the same site (and logging status) can be used. All plots of Greenheart forest are situated on the same soil type (Van Kekem *et al.* 1996), and plots of the same logging status are supposed to belong to the same homogeneous environment. If the species composition of plots is a random subset of all species suitable to grow within Greenheart forest, and given the shape of the distribution (Fig. 6.3), similarity in species composition between plots is expected to be relatively high. This can be illustrated by generating numerically random subsets of the species pool and calculating the Jaccard Index, *i.e.* the ratio of shared species and the combined species richness of plots. From a total of 18718 individuals divided over 340 species (the total number of measured individuals and species for all ground-rooted growth forms in Pibiri), 50 random subsets were drawn of 1250 individuals (mean number of measured number of individuals per plot). Then, the Jaccard Indices were calculated between pairs of subsets. It appeared that the Jaccard Indices varied between 0.82 and 0.97, because a few, very abundant species dominate the species composition of a subset. In reality, the Jaccard Index varied between 0.40 and 0.58 for all Pibiri plots, and the highest values varied between 0.54 and 0.58. For the unexploited plots highest similarity occurred mostly with plots in closest vicinity (table 6.2, table 4.6, Fig. 4.2), suggesting a distance dependency for the similarity of total species composition between plots. These results do not support the random walk hypothesis on the level of between plot variation in species composition, and would suggest that the species composition between plots is much more ruled by local processes such as seed availability.

On the level of within plot variation in species composition, in chapter 4 (table 4.7) it was shown that for most 24 one-ha plots of Greenheart forest there existed a significant positive correlation between the dissimilarity of subplots and their distance. This correlation was more often significant for the liana species than for the tree species. This would suggest that especially at finer spatial scales, the turnover of species (and in particular liana species) might be explained by geographical distance.

In conclusion the random walk theory can be criticized on three major points which make its testing impossible: 1) a trigger of a process may be random, but the subsequent processes are mainly deterministic. An example is gap formation, in which the occurrence of a tree - or branch fall may be a pure stochastic event, but the individuals that will establish -or proliferate- after gap formation are mainly determined by the surrounding vegetation. Thus a species composition is the result of both random and deterministic processes, even in an homogeneous environment. 2) a homogeneous environment is a theoretical notion which never can be established

Table 6.2
Jaccard Index of similarity between 15 one-ha plots of natural Greenheart forest in the West-Pibiri compartment, near Mabura Hill, Guyana.

All species Pibiri

	1	2	3	4	5	6	7	8	9	10	11	12	13	14
1														
2	**0.58**													
3	0.55	**0.56**												
4	0.55	0.51	0.52											
5	0.52	0.54	0.48	**0.56**										
6	0.48	0.48	0.45	0.47	**0.55**									
7	0.50	0.53	0.45	0.52	0.53	**0.54**								
8	0.51	0.51	0.47	0.48	0.54	0.54	0.50							
9	0.46	0.44	0.45	0.49	0.49	0.52	0.52	**0.56**						
10	0.51	0.51	0.48	0.51	0.53	0.51	**0.54**	0.54	0.55					
11	0.44	0.44	0.49	0.49	0.47	0.50	0.51	0.51	0.50	0.55				
12	0.47	0.47	0.46	0.40	0.48	0.51	0.47	0.53	0.47	**0.56**	**0.56**			
13	0.44	0.49	0.46	0.45	0.47	0.49	0.45	0.48	0.50	0.51	**0.58**	0.48		
14	0.50	0.47	0.50	0.50	0.51	0.53	0.48	0.54	0.52	0.55	0.54	0.50	0.54	
15	0.45	0.45	0.54	0.49	0.47	0.48	0.44	0.48	0.49	0.47	0.54	0.50	0.51	**0.58**

Liana species Pibiri

	1	2	3	4	5	6	7	8	9	10	11	12	13	14
1														
2	**0.58**													
3	**0.60**	0.57												
4	**0.54**	0.52	0.45											
5	0.49	**0.57**	0.42	0.51										
6	0.46	0.43	0.37	0.42	**0.57**									
7	0.46	0.49	0.37	0.46	0.56	**0.65**								
8	0.51	0.46	0.38	0.47	0.55	0.52	0.48							
9	0.49	0.41	0.45	0.46	0.49	0.55	0.50	**0.61**						
10	0.51	0.47	0.45	0.48	0.51	0.53	0.54	0.53	0.54					
11	0.44	0.42	0.41	0.46	0.50	0.56	0.51	0.54	0.51	0.59				
12	0.47	0.46	0.42	0.35	0.53	0.52	0.44	0.51	0.49	**0.60**	0.54			
13	0.45	0.44	0.44	0.43	0.52	0.48	0.41	0.45	0.52	0.50	**0.56**	0.47		
14	0.52	0.43	0.43	0.46	0.52	0.51	0.47	0.52	0.55	**0.57**	0.56	0.56	0.53	
15	0.51	0.43	0.51	0.49	0.53	0.58	0.47	0.47	0.54	0.54	**0.60**	0.53	0.55	0.56

Table 6.2 (continued)
Similarity values are given for all species, and separately for three growth-forms. **Bold** values are highest values for each plot.

Tree species						Pibiri								
	1	2	3	4	5	6	7	8	9	10	11	12	13	14
1														
2	**0.55**													
3	0.50	0.53												
4	0.51	0.49	0.55											
5	0.53	0.52	0.50	**0.59**										
6	0.50	0.52	0.50	0.52	**0.55**									
7	0.53	0.52	0.46	0.52	0.51	0.49								
8	0.49	0.54	0.51	0.47	0.54	0.53	0.51							
9	0.43	0.46	0.44	0.51	0.51	0.47	**0.53**	0.49						
10	0.49	0.52	0.48	0.49	0.54	0.49	**0.53**	0.50	0.52					
11	0.45	0.46	0.52	0.53	0.47	0.43	0.51	0.47	0.45	0.52				
12	0.45	0.45	0.48	0.41	0.45	0.48	0.48	0.51	0.41	0.50	**0.54**			
13	0.42	0.54	0.46	0.46	0.46	0.45	0.47	0.47	0.45	0.48	**0.58**	0.45		
14	0.49	0.51	0.52	0.54	0.50	0.53	0.50	**0.53**	0.51	0.52	0.52	0.47	0.52	
15	0.44	0.49	**0.58**	0.53	0.47	0.46	0.46	0.50	0.47	0.47	0.51	0.49	0.49	0.52

Undergrowth species						Pibiri								
	1	2	3	4	5	6	7	8	9	10	11	12	13	14
1														
2	**0.68**													
3	0.65	0.62												
4	**0.73**	0.63	0.67											
5	0.56	0.54	**0.64**	0.58										
6	0.42	0.40	0.48	0.38	0.48									
7	0.54	0.64	**0.68**	**0.75**	0.54	0.46								
8	0.65	0.55	0.67	0.59	0.50	0.63	0.54							
9	0.48	0.46	0.55	0.50	0.42	**0.68**	0.52	0.63						
10	0.61	0.58	0.55	0.70	0.54	0.59	0.58	**0.79**	0.67					
11	0.40	0.44	0.60	0.42	0.41	0.65	0.50	0.60	**0.74**	0.57				
12	0.61	0.58	0.48	0.56	0.48	0.67	0.52	0.70	**0.75**	0.81	0.64			
13	0.50	0.43	0.53	0.48	0.40	0.67	0.50	**0.71**	0.67	0.65	0.63	0.65		
14	0.52	0.43	0.61	0.48	0.52	0.58	0.44	**0.71**	0.50	0.57	0.55	0.50	0.65	
15	0.33	0.25	0.40	0.30	0.29	0.29	0.27	0.40	0.38	0.32	0.35	0.32	0.43	0.22

unambiguously, in reality the condition of a homogeneous environment can almost never be fulfilled, except at the smallest spatial scales. It remains the question if we are then not merely considering the niches of individual plants, and the results are trivial. 3) the pattern resulting from an analysis is not only related to the scale of observation, but also to the density and layout of the observations. Although processes underlying the pattern may be deterministic, we may not observe them simply because our observations do not permit us to. Thus we may perceive a patterns as random, even when it is caused by deterministic processes. The random walk may sometimes be the best way to describe a pattern at a given scale, although it says nothing about the nature of the processes causing the pattern.

On a large time scale (evolutionary time) the stability - time hypothesis might be an explanation for the diversity levels of tropical rain forest of the Guianas. On a smaller time scale (ecological time) we only find support for the gap partitioning theory for a small part of the species. Our data indicate that it is unlikely that one and the same theory suffices to explains the pattern in the different growth forms, as the documented patterns differ substantially between different growth forms and species groups. Geographical distance does not play a role in explaining differences in epiphytic species composition or the composition of undergrowth species, but it is the dominant factor in explaining differences in the species composition of lianas (at different spatial scales) and to a lesser extent for differences in species composition of trees. Epiphytes seem to be strongly determined by their environment. The simple notion 'everything is everywhere and the environment selects' seems to hold here. No complication due to seed availability or successional status seems to interfere with this pattern. On a very small spatial scale not every species can occupy the same space and at that (trivial) scale random sampling may be assumed. The clear geographical patterns that emerge for the lianas do not support the hypotheses that liana species composition is either purely determined by the mosaic of successional stages of the forest (gap dynamics theory) or a random subset of a larger pool of species (random walk theory). Although some liana species did show clear correlations with the mosaic of succesional stages, species composition was mainly determined locally.

Equilibrium theories view tropical community as a balanced community of many species whose coexistence is explained based on Lotka-Volterra competition and predation equations for species that require the same resources. This notion is still based on very simplistic models (Lotka Volterra competition in well mixed systems). Studying similar models which explicitely include space has revealed the likelihood of spatial heterogenous systems (Smith & Huston 1989; Hogeweg 1994; Van der Laan & Hogeweg 1995; Savill *et al.* 1997; Deutschman *et al.* 1997). In these models, local spatial interactions between individuals have far-reaching consequences for ecological systems because of the ability of many systems to self-organize into spatial patterns. The performance of an individual plant is based on the individual's local neighborhood. Spatial heterogeneity results from enhancement of initially very small (random) differences. Nevertheless, to refer to such spatial patterns as 'random walk' is not correct: the spatial pattern may be the result of deterministic processes. For instance, Deutschman *et al.* (1997) use an empirically based complex model of forest dynamics (SORTIE) and show that in this model forest dynamics (and levels of diversity) emerged as the result spatial interactions between individuals (local competition for light). In this model differences in responses of plants to their local light environment and the local dispersal of seedlings, both deterministic processes, leads to spatial heterogeneity. Excluding space (and spatial interactions) from the model resulted in lower levels of diversity (Deutschman *et al.* 1997).

Our data did not permit the evaluation of interactions between individuals (including compensatory mortality hypothesis) as an important process to maintain a high level of diversity in space. This mechanism of interaction, however, seems to be the only mechanism which explicitly incorporates spatial variation at different scales. Further research is needed to test if these interactions can explain at the same time the different responses of plants to logging and the relative unimportance of the mosaic. In particular, data are needed on shared characteristics of certain functional groups in relation to logging, *e.g.* plant physiological responses to light and water, growth rates in general, allocation patterns, and dispersal syndromes. Only for a few species occurring in Greenheart forest those data are available, and mainly for trees. The hegemony of trees in forest ecology research disregards the fact that competitive interactions on the forest floor occur between all ground-rooted individuals. It is time for more research regarding non-trees.

Acknowledgements

Discussions with Marinus Werger, Paulien Hogeweg, Roderick Zagt and Victor Jetten shaped the thoughts put forward in this chapter, and they commented on a draft version. Andre Noest provided the algorithms for the light zone calculations. Many thanks to all.

Appendix 6.1

Increased light level zonation

(*A.J. Noest*, Bioinformatics Group, Utrecht University)

Geometry

The hypothetical light distribution of the forest plots was estimated based on the map of canopy gaps, produced by PC-Raster. The basic idea underlying an effective approximation of the light distribution at ground level was to compute a year-averaged 'point-spread function $s(x,y)$, which is the light due to an imagined small element of a gap, at the origin $(x,y) = (0,0)$. The light in and around a set of arbitrary gaps can then be computed efficiently by convolution of the digitized map of gaps with $s\ (x,y)$.

Image a tiny gap positioned above the origin $(0,0)$. As seen from the ground at (x,y), the elevation-angle (α) and azimuth (β) of the gap-element are given by

(1) $$\sin\alpha = \frac{1}{x^2+y^2+1}$$

(2) $$\cos\alpha = \sqrt{\frac{x^2+y^2}{x^2+y^2+1}}$$

(3) $$\cos\beta = \frac{y}{\sqrt{x^2+y^2}}$$

when the x,y coordinates are measured in units of the canopy height.

The yearly amount of light falling through through this gap element depends on δ, the declination angle of the gap element when projected onto the celestial sphere. Using the classical transformation,

(4) $$\sin\delta = \sin\alpha\ \sin\phi - \cos\alpha\ \cos\phi\ \cos\beta$$

one obtains,

(5) $$\delta = \arcsin\left[\frac{\sin\phi - y\ \cos\phi}{\sqrt{x^2+y^2+1}}\right]$$

where ϕ is the nautical longitude of the observer.

Year-averaged point-spread function

To find the year-mean point-spread function *s(x,y)* caused by our imagined small gap element at *(0,0)*, one need to multiply (a) the brightness distribution (year-average) of solar light on the celestial sphere, (b) the atmospheric transmission factor, (c) the reduction of the effective cross-section of the gap-element, and (d) the reduction in light intensity per unit of ground surface for lower elevations. Finally, (e) the interception of light by the surrounding vegetation (mainly treetrunks) should be accounted for. Before tackling the main points one by one, it can be noted that points (c) and (d) each simply give rise to a factor of $\sin\alpha$. Together this implies an *x,y*-weighting factor $(x^2 + y^2 + 1)^{-1}$.

a) Averaged solar brightness factor

The solar δ goes through one cycle per year, between $\pm D$. Using t as time since spring-equinox (in years), we may write $\delta = D \sin(t)$, with $D = 23.645$ degrees.
The number of days in a year is sufficient large for treating the year distribution $p(\delta)$ of solar δ realized during day-time as if it were continuous, except at the solstitia (where $\delta = D$).

(6) $$p(\delta) = \frac{1}{\pi\partial_t\delta} = \frac{1}{\pi D\cos(t)} = \frac{1}{\pi D\sqrt{1-(\delta/D)^2}}$$

To avoid numerical problems with the sharp cusps of $p(\delta)$ at the solstitia, we actually calculated the mean value over an interval of width 2ρ (say 1 degree), centered on the variable δ. Thus we get

(7) $$p_\rho(\delta) \equiv \frac{1}{2}\rho \int_{\delta-\rho}^{\delta+\rho} dx \frac{1}{\pi D\sqrt{1-(x/D)^2}} = \frac{\arcsin*[\frac{\delta+\rho}{D}] - \arcsin[\frac{\delta-\rho}{D}]}{2\pi\rho}$$

in which an extended arcsin-function is used, defined as

(8) $$\arcsin^*(x) = \arcsin(x); |x| < 1$$

(9) $$= \frac{\pi}{2}\, sign(x); |x| \geq 1$$

Finally, we have to account for the fact that the sun travels slower along its daily track for larger values of $|\delta|$. Per day, its track in celestial coordinates is very nearly a circumpolar circle with circumference $2\pi\cos\delta$. Thus, we obtain the year-mean brightness distribution across the declinations as

(10) $$B(\delta) = \frac{\arcsin^*[\frac{\delta+\rho}{D}] - \arcsin^*[\frac{\delta-\rho}{D}]}{4\pi^2\rho\cos\delta}$$

Using eq. 5, the δ-dependence may be expressed in terms of the coordinates x,y and, therefore, we may re-use the same function to write the brightness factor in the point-spread function as $B(x,y)$.

b) Atmospheric transmission factor

Even an cloudless atmosphere reduces the amount of light reaching the surface, because of a variety of scatttering and absorption mechanisms. Again, we resort to a rough approximation. First, one introduces the effective transmissionfactor $e^{-\tau}$ for a vertical ($\alpha = \pi/2$) light path. Typically, $e^{-\tau} = 0.8$ at sea level. Next, we must account for the longer pathlength at low elevations α. We neglected the curvature of the earth and the vertical structure of the atmosphere. The relative effective pathlength is then simply $\sqrt{(x^2 + y^2 + 1)}$, and the transmission factor becomes

(11) $$\exp(-\tau\sqrt{x^2+y^2+1})$$

c) Interception of light by the surrounding vegetation

Accounting accurately for the interception of light by the surrounding vegetation outside a gap was not possible, so we resorted to computing an upper and a lower bound.
The upper bound simply neglects the interception of light by the surrounding vegetation and would be a valid approximation for inside the gaps. Combining the previous mentioned equations gives the following point-spread function

(12) $$\hat{s}\,(x,y) = \frac{B(x,y)\ \exp\ (-\tau\sqrt{x^2+y^2+1})}{x^2+y^2+1}$$

To find the lower bound for the interception of light by the surrounding vegetation we accounted for the (on average exponential) extinction of light as it is intercepted randomly by the trunks of trees, which are mostly vertically oriented. This produces a lowerbound point-spread function, valid for the situation outside gaps, of the form

(13) $$\check{s}\,(x,y) = \hat{s}\,(x,y)\ \exp\ (-\eta\sqrt{x^2+y^2})$$

Thus we need to estimate η. We assumed that interception is dominated by treetrunks, at least in the region below the canopy. This enables us to use available data on $p(d)$, the number of stems of various diameters d, the density of trees of different diamter classes within the one-ha plots. In fact, we found that the 'penetration depth' of light so computed ($1/\eta$) is about 1.5 times the canopy height. To account for the addition interception by branches we have actually used $\eta = 1$ in our calculations.

Computing the increased light levels in and around gaps

We can pre-compute the upper and lower bound point-spread functions, and save a factor of two because they are *x*-symmetrical. Next, we represent the map of gaps as a binary valued function *g(x,y)*, which equals 1/0 at points inside/outside a gap. The respective bounds of lightdistribution *L(x,y)* for this map are then formally given by an convolution integral of the form

$$L(x,y) = \int du \int dv \; g(u,v) s(x-u, y-v) \tag{14}$$

Using the two bounds $\hat{s}(x,y)$ and $\check{s}(x,y)$, we obtain the corresponding upper and lower bounds for *L (x,y)*. These integrals are actually computed via Fourier transformations, which reduces the numerical complexity for an N-point map from $O(N^4)$ to $O(N^2 \log N)$ (Press & Flannery 1988).

The final map of light distribution over the plots is obtained by using the computed light levels for the upper bound limits, without interception by surrounding vegetation, for inside gaps and the computed light levels with the lower bound limits for the areas outside the gaps.

CHAPTER 7

Conclusions and recommendations for forest management

One of the objectives of this study was to contribute to a sustainable management plan for the Greenheart forests of Guyana. Sustainable management implies a balance between the economical and social benefits of logging and resulting logging damage. In relation to the biodiversity of Greenheart forest it would mean a balance between the amount of commercial timber harvested and the extent of change of the characteristic species composition of the forest due to logging. But what is exactly that characteristic species composition of Greenheart forest?

In this thesis the effects of logging on the species composition of Greenheart forest are evaluated. Data on the plant species composition and their abundance in unexploited Greenheart forest were gathered, together with data on the plant species composition at different time steps after logging. This thesis, therefore, focusses on ecological contributions to a viable management system.

The conclusions and recommendations formulated in this chapter are mainly based on the data from conventionally logged areas, Waraputa, Reserve, and '2km', and unexploited Greenheart forest from 15 one-ha plots in a reduced-impact logging project area, Pibiri, prior to logging. Only preliminary results, from two re-enumerated Pibiri plots two years after logging and only for the lianas, are included for the reduced-impact logging project.

Species Diversity

Greenheart forest was found to be a mediate species-rich to species-poor forest type, compared to other mixed tropical rain forest types in the Neotropics. High species diversity in Greenheart forest could be found in both unexploited and exploited forest (table 4.3, 5.1). Although within the same site species richness increased with logging, number of species alone was not an appropriate measure for forest disturbance. The species composition before and after logging was significantly different (table 4.8, 5.6). Unlogged plots had a different floristic composition, with species that prefer, or can endure, lower light levels, while in the logged areas species with preferences for higher light levels become dominant (table 5.5, group 1).

* ***Species richness, number of species in a specified area, is not an appropriate measure for forest disturbance.***

Species diversity of lianas was positively correlated with tree species diversity, and total species diversity, both for unlogged and logged plots (chapter 4). Thus, either of the two can be used to obtain a reliable estimate of the total species richness. Furthermore, the number of liana species after logging showed the highest correlation with the extent of change in the species composition. Species diversity of undergrowth species (herbs, shrubs, and small palms), however, was only slightly correlated with total species diversity, and only in unlogged forest (table 4.5). Although far less undergrowth species occurred than tree or liana species, the undergrowth species have to be explicitly included in sampling to obtain reliable estimates on the species richness and abundance of this growth form. The undergrowth species showed a much finer-grained mosaic of occurrences compared with the trees and lianas, and can be sampled at a smaller spatial scale.

Six and ten years after logging the effects of extraction are still apparent in a changed species composition and the abundances of the species concerned. There are no indications that there is a decrease in species richness or abundance at the time scale of ten years (chapter 4, 5).

The effects of logging, *i.e.*, the highest number of species that established after logging only, were noticeable in different size classes for the different growth forms (table 4.10). For trees the effects are mostly visible within the size class of trees with a diameter at breast height between 1 and 5 cm, for lianas within the group of lianas with a height of more than 2 m, and for the herbs and shrubs in all size classes. Two years after logging, the effects of logging are visible in the group of lianas with a height between 50 cm and 2 meters (chapter 5). Different size classes were enumerated in different size subplots, *e.g.*, trees with a dbh between 1 and 5 cm were enumerated in 5 by 5 m subplots, and lianas with a height more than 2 meters were enumerated in 10 by 10 m subplots. This gives a much larger and more reliable data set for the latter growth form. It is recommended for future enumerations to use larger sampling plots for those size classes of individuals that are expected to have established shortly after logging, and to situate those sampling plots in the various areas that experienced forms of damage as a result of the logging activities.

Logging Damage

Logging damage or logging-induced changes in the environment were found to lead to clear patterns of diversity. The new tree species that established after logging (chapter 4), for instance, showed a clear correlation with the extent of overlapping gap and skid trail area. This area showed also a correlation with the new liana species that established after logging. Consequently, the effect of skidding within a gap had the most pronounced effect on the total species composition.

For the trees the number of species after logging was mainly dependent on the species that established after logging. Most tree species that were present before the logging were still present after logging, although their relative abundances changed. The number of liana species after logging (chapter 5) was positively correlated with the number of species present before logging (previous occupation and opportunities of seed dispersal), and negatively correlated with the amount of logging damage, expressed in number of harvested trees and skid trail area. These results indicate that logging will always lead to an increase of liana species, as a consequence of creating new establishment sites. Moderate logging intensities will result in a mix of both disturbed and undisturbed forest patches. The more species rich the forest is before logging, the more species will have a chance to establish themselves after logging. After logging, especially the number of common liana species increased, compared to rare liana species (chapter 5). Higher logging intensities, *i.e.* increased numbers of harvested trees per area, will lead to an increased area that is heavily disturbed and to a decreased area of undisturbed forest. Consequently, with higher logging intensities more common species will appear, at the cost of rare species, and the species richness after logging will decrease.

* ***Number of harvested trees per area showed a correlation with the number of liana species per area after logging. The number of liana species after logging in its turn showed the highest correlation with the extent of change in the species composition. The number of harvested trees per area should be kept as low as possible***

.... and in particular gaps

As expected, conventional logging caused larger gap-areas (Reserve on average 33.8% of the total area, Waraputa 35.9%, 2km 37.8%) than natural tree gaps (Reserve 9.7% of the total area, Waraputa 16.6%, '2 km' 17.4%) (table 7.1). These figures give the total of directly affected area inside gaps, where gaps are measured as a vertical projection of a canopy opening enclosed by the crown edges of the surrounding trees. The actual area influenced by the gaps, and specifically by increased light levels, was much larger. Forest areas outside gaps, with increased light levels but no surface disruption, were richer in number and abundance of species (table 5.5). The extent of the area outside a gap with increased numbers of species and individuals was dependent on gap size and orientation, and the height of the surrounding canopy (chapter 4).

The number of new, invading species after logging (all growth forms) showed correlations with the amount of gap area, and in particular large gap area.

Gap size, and orientation, has an influence on almost all processes that regulate biodiversity, both biotic and abiotic (Ter Steege *et al.* 1996). The larger the gap, the larger the extent of changes in microclimate, compared to the forest understorey. Gap partitioning between small and large gaps seems to play a role (chapter 6), with some species showing a distinct preference for a specific gap size, or light environment. Most species, however, show no such clear preferences. The composition of the species that fill a gap is mostly locally determined, and the species composition in and around small gaps has the highest similarity with the species composition in the remaining 'undisturbed' forest patches. This composition was, in general, significantly different from that of large gaps in the same area (chapter 4). Around small gaps the surrounding forest may be able to buffer the impact of the logging by lateral ingrowth of the surrounding canopy (Van der Meer & Bongers 1996b), decreasing the amount of nutrient leaching (by root activity), and reducing the nutrient losses through litter input (Brouwer 1996): environmental parameters such as amount of nutrients, radiation, air temperature, humidity, soil temperature and water content in small gaps have the highest similarity with those of unexploited forest.

* ***The species composition in and around small gaps had a higher similarity with the species composition of undisturbed forest patches, compared to the species composition in and around large gaps. Gap size should be kept as small as possible.***

Brouwer (1996) suggests that creating a series of small gaps is less disturbing for the nutrient cycle than creating one single large gap. He states that the 'allowable' gap size, in which leaching losses are modest and the boundary effect of surrounding forest is large, is probably about 500 m^2, which presents a best guess'. In the present 'Botanical Diversity' study small gaps are defined as comprising less than 300 m^2 and large gaps as comprising equal to, or more than, 300 m^2. This distinction results in significant differences in species composition (table 4.8). I would suggest, therefore, to limit the 'allowable' gap size to 300 m^2. Please note that this size limit is 'inspired' by the (arbitrary) size limit used in this study. In a current study on the effects of different intensities of logging P. Van der Hout (pers. com.) found that in practice it is indeed possible to log the forest without creating gaps of more then 300 m^2. The ongoing research of the Tropenbos Guyana Programme on the influence of gapsize on the regeneration of the forest, should give more precise indications on 'allowable' gap sizes.

* ***The 'allowable' gap size, in which the species composition is not significantly different from unexploited forest, is 300 m^2.***

That *Cecropia* is an indicator of large disturbed areas comes as no surprise. Everyone who ever visited the Neotropics will agree. For the trees, however, it was the only species that showed a significant positive correlation with large gaps (table 4.13). Large gaps were furthermore characterized by the presence of the lianas *Coccoloba parimensis, Machaerium myrianthum* (Hold-me-back, Wait-a-second), *Maripa scandens* (Monkey syrup), *Memora racemosa, Pinzona coriacea* (Fire rope), and *Passiflora* spp. (Passionfruit). From the undergrowth species, especially the shrubs/treelets *Miconia gratissima* (Waraia) and *Miconia punctata* (Waraia) were common in large gaps, and sometimes the grass *Scleria* sp. (Razor grass). Six and ten years after logging these species are still present, even after the original size of the gaps diminished. The presence of those species indicates the presence of a former large gap.

Small gaps were positively correlated with only a few species: the tree species *Annona symphyocarpa* (Duru) and *Aspidosperma exselsum* (Yaruru), and the liana species *Curarea candicans* (a kind of Devildoer).

.... and in particular skid trails

Skid trail areas varied between 15-20% in conventionally logged areas (Reserve 17.8 %, Waraputa 16.5%, 2km 19.2%) (table 1). The total affected area by gaps and skid trails is 43.9% for the Reserve area, 43.1% for Waraputa and 47.3% for '2 km' (table 7.1).

As stated before, the amount of overlapping gap and skid trail area, together with both areas separately, are the most important factors that correlated with number of species after logging and number of new, invading species after logging.

* ***total skid trail area has a significant influence on number of species after logging; skid trail area should be kept to a minimum.***

* ***combined gap and skid trail area has the strongest influence on the number of newly established species after logging; skidding within gaps should be kept to a minimum.***

Table 7.1
Extent of logging damage in the traditionaly logged sites.

	Unlogged		Logged1		Logged2	
	m2	%	m2	%	m2	%
Waraputa (logged in 1989-1990)						
gaps	1655	16.6	3615	36.2	3569	35.7
skid trail			1692	16.9	1613	16.1
total for gaps & skid trail			4224	42.2	4401	44.0
interior	8345	83.4	5778	57.8	5599	56.0
number of felled trees			16		25	
Reserve (logged in 1989)						
gaps	1066	10.7	2827	28.3	3949	39.5
skid trail			1835	18.4	1732	17.3
total for gaps & skid trail			4007	40.1	4788	47.9
interior	8934	89.3	5993	59.9	5213	52.1
number of felled trees			20		21	
2 km (logged in 1985)						
gaps	1739	17.4	4112	41.1	3455	34.6
skid trail			2216	22.2	1632	16.3
total for gaps & skid trail			4758	47.6	4708	47.1
interior	8261	82.6	5242	52.4	5292	52.9
number of felled trees			29		22	

Skidder activities will lead to soil compaction, uprooting of the soil and damage to the remaining stand (Ter Steege *et al.*1996; Bruijnzeel & Critchley 1994; Van der Hout 1996), all together impacts dissimilar to those of any natural disturbance. Although the canopy over skid trails may remain closed, it is at least thinned out at lower levels, leading to higher light availability at the forest floor. Both the compacted and uprooted soil forms a new habitat for germination and establishment for species, due to *e.g.*, changes in water retention capacity (Jetten 1994), and short-term changes in nutrient availability by decomposition of the overrun vegetation (after Brouwer 1996). However, Pinard *et al.* (1996) consider skid trails an unfavorable habitat for establishment. Furthermore, Brouwer (1996) found increased (possible toxic) Aluminum levels on skid trails. Species found growing on skid trails must therefore be tolerant to these unfavorable site conditions and seem favoured by the uprooting of the soil, and the increased levels of light and water availability.

Species that showed significantly positive correlations in number of individuals with increased area of skid trails were found in all growth forms (table 4.13). For the trees especially *Emmotum fagifolium* (Manobodin) correlated both significantly with skid trail and with the combined gap and skid trail area. Furthermore, *Goupia glabra* (Kabukalli), correlated with this combined area. For the lianas the strongest correlation was found with *Passiflora* spp. (Passionfruit species), and with Fire rope (*Pinzona coriacea*), also for the combined gap and skid trail area. Within the group of the undergrowth species, the principal species growing on skid trails is Waraia (*Miconia gratissima*) and another unidentified Waraia (*Miconia* sp.). Two other species, the Inkberry (*Renealmia orinocensis*) and a kind of Mucru (*Ischnosiphon gracilis*) correlated strongly with the combined gap and skid trail area.

Again, all these species are still present six and ten years after logging and can assist in the identification of these different forest zones in previously logged sites

.... and again: lianas

Already many remarks were made concerning lianas. Lianas, though, did show the highest correlations with logging damage. What makes this growth form special in relation to logging is the fact that foresters consider lianas as a nuisance and pre-harvest liana cutting is part of many silvicultural treatments (Fox 1968; Appanah & Putz 1984; Putz 1991; Van der Hout 1996; Mason 1996). Liana cutting was also undertaken within the reduced-impact logging project, three to seven months prior to logging (Van der Hout 1996). The effects of this liana cutting have to be evaluated in a later phase, after completion of a total re-enumeration of Pibiri.

Lianas were found to be occasional to dominant in the colonizing vegetation of gaps and skid trails, sometimes leading to impenetrable liana vegetations ('liana blankets') after logging. One and a half year after logging part of the large gaps in Pibiri were found to contain very many, mostly creeping, liana individuals. At that moment they did not form a real blanket, but one year later they did (Van der Hout, pers. comm.). Only a small part of the liana species are able to form such a 'liana blanket'. The establishment of commercial species can be hampered by such a liana blanket. Post-harvest silvicultural treatments, in which the liana blanket is (partly) removed, are recommended to ensure future commercial potential of the forest. However, this measure is not needed when gap size is kept small, and liana blankets will not occur.

* ***To prevent some liana species to form a 'liana blanket', limited post-harvest silvicultural treatments are recommended.***

Lianas showed the highest relative increase in species numbers and after logging this growth form formed a larger portion of the total species pool. It is therefore expected that in areas that are repeatedly disturbed, lianas will become the dominant growth form and will form a 'ropy bush' or liana forest.

A group of lianas can be distinguished which is able to connect more host-tree species and which can reach a diameter of more than 5 cm. This group of lianas could have a higher impact on logging damage.

* *Only some of the liana species are able to connect more host-trees and develop such a diameter that they can have a high impact on logging damage.*

* *Pre-harvest liana cutting only needs to be applied to those liana species able to connect more host-trees. A small, simple, identification key for those species would be needed to facilitate the recognition of those species.*

Four (4) different liana groups could be identified, each with a different response to logging:

1 Invading species, mainly appearing after logging and mostly establishing in large gaps and on skid trails.
2A Species indifferent to logging, with a preference for higher light levels, and mostly found around gap edges, including natural gaps.
2B Species indifferent to logging, with a preference for lower light levels, mostly found at a distance of at least 10 meters from a gap edge.
3 Exploitation sensitive species, regenerating in undisturbed forest patches and disappearing or declining in numbers after logging.

These results concerning lianas, and their correlations with other growth forms, are formulated on the basis of data from the older logged areas and the two re-enumerated Pibiri plots. For the older logged plots an unlogged referential plot within the same area served as representing the former unlogged status of those plots. To validate whether the correlations between lianas and other growth forms are still applicable when the pre-logging botanical diversity is known, a larger amount of re-enumerated plots is necessary. I hypothesize that: 1) the four liana groups can be considered as functional groups which possess similar characteristics and adaptations to a changing environment; and 2) the responses of the functional groups of lianas to logging are similar to the responses of other growth forms in the forest to logging. If this last hypothesis is valid, the liana species composition can be used to assess the impact of logging damage on the biodiversity.

Most studies monitoring the reactions of the forest ecosystem to disturbance (logging) emphasize on the tree component of the forest (*e.g.*, Hawthorne 1993; Bruenig 1996). However, the liana vegetation shows logging-induced diversity patterns earlier in time (chapter 4 & 5). Within the group of lianas different functional groups could be distinguished comparable to those already described for trees (*e.g.*, Clark & Clark 1990; Denslow 1996) (chapter 5). Growth rates of lianas are hardly known (Putz & Mooney 1990), but are expected to be higher than those of trees. Lianas, therefore, could be used to assess the effects of logging damage in a shorter time-span than required when studying this on trees.

Future research is needed to evaluate the effects of reduced-impact logging, with different logging intensities.

CHAPTER 8

Summary

In this thesis the impact of disturbance, and in particular logging, on species diversity and functional diversity of Greenheart (*Chlorocardium rodiei*) dominated rain forest in Guyana is evaluated. The main goal is to assess the change in the composition and development of the botanical community in Greenheart forest as a result of logging and logging-related damage. Data on the plant species composition and their abundance in unexploited Greenheart forest were gathered, together with data on the plant species composition at different time steps after logging. The effects of conventional logging (with high logging intensities and damage) on the plant species composition and abundances were evaluated, with special reference to the influence of skid trails and gap sizes. In this study all growth forms were considered: herbs, shrubs, palms, lianas, epiphytes, and trees. Other objectives were to evaluate various hypotheses on the maintenance of diversity and to contribute to a sustainable management plan for this forest type containing Guyana's main commercial timber species.

From a theoretical point of view, the observed levels of diversity can be explained by different theories. These are discussed in **chapter 6**. From a practical point of view, the implications of changing biodiversity for forest management are discussed in **chapter 7**.

In **Chapter 1** a general introduction to biodiversity in tropical rain forests, and the possible influence of logging, is presented. The different meanings of the term 'biodiversity' are discussed, and the objectives and approach of this study are outlined.

In **Chapter 2** the regional and geographical setting of the flora of the Mabura Hill area is discussed. The flora of the Mabura Hill area is described and compared with the flora of other Neotropical sites, and in particular with the flora of other sites from the Guayana Shield. The flora of the Mabura Hill area contains mainly primary forest elements, with an apparent poverty of undergrowth (herbs, shrubs, and small palms) and epiphytic species, especially when compared to two French Guianan sites. The tropical rain forest near Mabura Hill has a mediate to low species diversity and hardly any introduced or 'weedy' species, which indicates little large-scale natural disturbance. In this chapter it is argued that the origin of the most important families in the Mabura Hill area, the large proportion of Caesalpinoideae in the Guayana Shield and Northern Brazil sites, and the high percentage of such families as Connaraceae, Menispermaceae, Nyctaginaceae, Annonaceae, and Apocynaceae support the hypothesis that the flora of the Guianas shows a strong link with the forest flora of Western Africa, stronger than with other parts of South America. This pattern might be explained by the low disturbance status of the area.

Chapter 3 focusses on the diversity of one particular growth form within Greenheart forest, the epiphytes. Its patterns of occurrence are evaluated in a decreasing scale, from site specific, via forest type specific to host type specific. The epiphytes found in Greenheart forest were therefore compared with epiphytes occurring within three other forest types, both in Guyana and French Guiana. Each forest type harboured its own site specific epiphytic community, with Greenheart forest as the forest type with lowest species numbers. In all four forest types the same few species were found dominating the epiphytic vegetation. Epiphytic species that occurred in more than three forest types showed comparable vertical distribution patterns in each forest type. Similarities of the four forest types were mainly found within the canopy zones, and these zones also showed the highest species richness. Within each forest type, four groups of major epiphytic habitats, each with its own functional group of epiphytic species, could be distinguished. Geographical distance did not explain differences in species richness between forest types. Humidity conditions and host specific factors, such as facilitation of the bark and tree architecture, gave higher correlations. The species composition of epiphytes seems to be largely determined by their environment.

In **Chapter 4** a description is given of the species composition and abundance of unexploited Greenheart forest. The low levels of species diversity of the total flora of the area (chapter 2) and epiphytes within this forest type (chapter 3) were also found for the total diversity of Greenheart forest. Geographical distance was a very dominant factor in explaining differences in species composition of lianas, and to a lesser extent of trees, but not of undergrowth species. The relation between changes in species diversity and logging damage as a result of conventional logging methods was evaluated. All ground-rooted growth forms showed their own responses to logging, at different spatial scales. Lianas were the most responsive to logging-induced changes in the environment and showed the highest relative increase in species number. The composition of tree species after logging was mainly determined by newly established species after logging. The more heavily damaged an area became during logging, such as areas where gap and skid trail overlap, the more new, invading, species arrived after logging (both trees and lianas). Undergrowth species (herbs, shrubs and small palms) were very uniform in species composition over exploited and unexploited forest, with a strong dominance of a few species. The species composition in large and small gaps appeared to be mainly determined by the surrounding vegetation. Six and ten years after logging the effects of extraction were still evident in a changed species composition and their abundances. There were no indications that, on a one hectare scale, a decrease in species richness or abundance can be found at the time scale of ten years.

In **Chapter 5** the responses of lianas to logging are evaluated. The composition of the liana community (in species richness, diversity, and abundance) was correlated to the extension of different logging-related damage zones. The lianas constituted the only growth form for which the exact species composition and abundance on a particular location were available before and after logging. This allowed much more detailed evaluation of the responses of this growth form to logging. In particular, the spatial patterns of those individuals which established shortly after logging were compared to different definitions of damage zones. Clear patterns of liana diversity that resulted from logging, or logging-induced changes in the environment, were documented. Number of rare species of all growth forms together showed no differences between unexploited and exploited Greenheart forest. The lianas, however, clearly showed an increased number of common species compared to rare species after logging. Unexploited Greenheart forest had a different liana vegetation composition, with lianas that prefer or can endure lower light levels. In the exploited sites species with preferences for higher light levels became dominant. Based on the behavior of individual species four different functional liana groups could be defined, each with its own response to logging. Some liana species clearly declined in numbers or disappeared completely as a result of logging, some reacted indifferently to logging and others proliferated after logging. The species indifferent to logging could be divided in a group of species which showed clear preferences for higher light levels and a group with preferences for lower light levels. Turnover of species and individuals for two plots enumerated before and two years after logging, showed that especially the area outside gaps, with increased light levels but no surface disruption, had favourable conditions for establishment of liana seedlings shortly after logging. These outer gap zones harboured species already present before the logging as well as newly arriving species. Six and ten years after logging this increase in number of species and individuals was not found, suggesting that especially in these 'undisturbed' outer gap zones, number of species and individuals decreased at the time scale considered. These changes occur at finer spatial scales and are not visible at the spatial scale of a whole ha (in contrast to chapter 4).

In **Chapter 6** the patterns of diversity found within Greenheart forest and the flora of the Mabura Hill area were used to evaluate the applicability of theories on the origin and maintenance of species diversity. From the available theories, the stability-time hypotheses may be able to explain on an evolutionary time scale, the present day diversity levels of tropical rain forest of the Guianas. On a finer 'ecological' time scale these hypotheses can be divided in two main groups, equilibrium and non-equilibrium theories. Equilibrium theories view certain species as competitively superior over all others within a specific combination of environmental conditions. To maintain high levels of diversity there should either be a very heterogenous environment or regular disturbances that will lead to a mosaic of forest patches in different stages of succession. In contrast, non-equilibrium theories view species as fairly similar in competitive abilities, and the likelihood of outcompeting other species rather small. Then, observed spatial variation reflects merely differences in randomly drawn samples from a large pool of species suited to grow in a given environment. The data indicate that it is unlikely that one and the same theory suffices to explain the pattern in different growth forms, as the documented patterns differ substantially between growth forms and species groups. Geographical distance does not explain differences in epiphytic species composition or the composition of undergrowth species, but geographical distance is the dominant factor in explaining differences in the species composition of lianas (over a range of spatial scales) and to a lesser extent for differences in species composition of trees. Epiphytes seem to be strongly determined by their environment. The clear geographical patterns that emerge for the lianas do not support the hypotheses that liana species composition is either purely determined by the mosaic of successional stages of the forest or a random subset of a larger pool of species. Only a few liana species did show clear preferences for specific forest zones and the major part of the liana species composition was determined locally.

Our data did not permit the evaluation of (frequency dependent) interactions between individuals (competition, including compensatory mortality hypothesis) as an important process to maintain a high level of diversity in space. This mechanism of interaction between individuals, however, seems to be the only mechanism which explicitly incorporates spatial variation at different scales and the ability of individuals to influence the structure and environmental conditions in which they occur. Further research (both in the field and theoretical) is needed to test if these interactions can explain at the same time the different responses of plants to logging and the relative unimportance of the mosaic. In particular data are needed on shared characteristics of certain functional groups in relation to logging, *e.g.* plant physiological responses to light and water, growth rates in general, allocation patterns, and dispersal syndromes. Only for a few species occurring in Greenheart forest those data are available, and mainly for trees. The hegemony of trees in forest ecology research disregards the fact that competitive interactions on the forest floor occur between all ground-rooted individuals. It is time for more research regarding non-trees.

In the future, when more data are available on the composition and abundance of species before and after logging on exactly the same location, more detailed evaluations concerning maintenance of species diversity theories will be possible.

Lastly, in **Chapter** 7, the observed changes in total species diversity and the diversity of different growth forms to logging were used to formulate conclusions and recommendations concerning forest management. The following recommendations were formulated:

Number of harvested trees per area showed a correlation with the number of liana species per area after logging. The number of liana species after logging in its turn showed the highest correlation with the extent of change in the species composition. The number of harvested trees per area should be kept as low as possible.

The species composition in and around small gaps had a higher similarity with the species composition of undisturbed forest patches, compared to the species composition in and around large gaps. Gap size should be kept as small as possible. The 'allowable' gap size, in which the species composition is not significantly different from unexploited forest, is 300 m^2.

Total skid trail area has a significant influence on number of species after logging; skid trail area should be kept to a minimum. Combined gap and skid trail area has the strongest influence on the number of newly established species after logging; skidding within gaps should be kept to a minimum.

To prevent some liana species to form a 'liana blanket', limited post-harvest silvicultural treatments are recommended. Only some of the liana species are able to connect more host-trees and develop such a diameter that they can have a high impact on logging damage. Pre-harvest liana cutting needs to be applied only to those liana species able to connect more host-trees. A small, simple, identification key for those species is needed to facilitate the recognition of those species.

Samenvatting

'Een goed begrip van de meest relevante aspecten van de biodiversiteit van tropische regenwouden, en het vermogen om de invloed van menselijke ingrepen op deze biodiversiteit te kunnen voorspellen, zijn essentieel voor elke poging de biologische rijkdom van deze tropische regenwouden te behouden en beheren'.

Bovenstaande uitspraak is te vinden in het beleidstuk van de Nederlandse Stichting Tropenbos aangaande prioritering van het onderzoek naar biodiversiteit (Tropenbos 1996). Deze uitspraak geeft de kern van het onderzoek zoals gepresenteerd in dit proefschrift weer. Tegelijkertijd introduceert deze uitspraak een aantal termen die veel gebruikt worden in relatie met biodiversiteit. Deze begrippen worden hier nader toegelicht.

Tropische regenwouden: Tropische regenwouden komen vooral voor in laaglandgebieden en heuvels tussen de Kreeftskeerkring en de Steenbokskeerkring, in die gebieden waar meer regen valt dan 1500 - 2000 mm per jaar. Tropische regenwouden bedekken maar 7% van het aardoppervlak maar herbergen wel naar schatting meer dan de helft van alle organismen die op aarde voorkomen. Tropische regenwouden werden in eerste instantie gezien als zeer stabiele en onveranderbare ecosystemen, waar de verschillende planten- en dierensoorten in evenwicht met elkaar voorkwamen. De laatste tijd wordt het tropische ecosysteem echter steeds meer als een dynamisch geheel beschouwd. De dynamiek van tropische regenwouden wordt veroorzaakt door verschillende vormen van verstoringen (bijvoorbeeld orkanen of branden) en periodiek variërende omgevingsfactoren (een voorbeeld is de huidige droogte in Indonesië, voortkomend uit een periodiek terugkomend verschijnsel 'El Niño' geheten). Hierdoor kunnen zowel kleinschalige en/of kortdurende effecten optreden als grootschalige en/of lange termijn effecten.

Invloed van menselijke ingrepen: Aanzienlijke delen van het tropisch regenwoud worden jaarlijks aangetast als gevolg van commerciële houtkap-activiteiten. Samen met het kappen en branden van tropisch regenwoud voor het creëren van landbouwgronden is de commerciële houtkap één van de grootste bedreigingen voor het voortbestaan van tropische bossen. Met het verdwijnen van de tropische bossen kunnen plant- en dierpopulaties verdwijnen. In extreme gevallen kan dit leiden tot het uitsterven van soorten. De publieke aandacht en internationale bezorgdheid rond dit thema heeft geleid tot een nieuw begrip: biodiversiteit.

Biodiversiteit: Het begrip 'biodiversiteit' wordt in verschillende vakgebieden gebruikt en heeft heel wat verschillende definities. In het algemeen zijn deze definities in drie grote categorieën samen te brengen: biodiversiteit als abstract idee, als meetbare eenheid en als sociaal of politiek concept. Als abstract idee wordt onder biodiversiteit verstaan: alle vormen van leven, variërend van moleculen en genen, via soorten, naar groepen van functioneel verwante soorten die in eenzelfde landschap voorkomen. Als meetbare eenheid wordt biodiversiteit meestal uitgedrukt in aantal soorten (soortenrijkdom). Dit verklaart waarom soortenrijkdom vaak als synoniem voor biodiversiteit wordt beschouwd. Diversiteit wordt echter beschreven door twee parameters: soortenrijkdom (aantal soorten per oppervlakte eenheid) en abundantie (aantal individuen per soort per oppervlakte eenheid). Tenslotte, als sociaal en politiek concept omvat biodiversiteit het effect op de mens van verandering in landgebruik, dat wil zeggen in hoeverre een verandering in soortensamenstelling gevolgen heeft voor de leefwereld van de mens, zowel fysiek als spiritueel.

De internationale publieke verontrusting over de teruggang in biodiversiteit, achteruitgang van bossen en de sociale ongelijkheid in de wereld heeft in 1992 geleid tot de Biodiversiteits Conferentie van Rio, de Conferentie over Milieu en Ontwikkeling van de Verenigde Naties (UNCED), gehouden in Rio de Janeiro. Deze conferentie heeft geleid tot een verdrag over Biologische Diversiteit. In dit verdrag werden een aantal aanbevelingen geformuleerd om een halt toe te roepen aan de vernietiging van soorten, leefmilieus en ecosystemen, genoemd 'Agenda 21'. Aanbevolen werd om internationaal te streven naar 'het behoud van biologische diversiteit, het duurzaam gebruik van deze diversiteit én de eerlijke en gelijkwaardige verdeling van de opbrengsten afkomstig van het gebruik'. In navolging van dit verdrag zijn er internationaal veel nieuwe initiatieven ontplooid en zijn er beleidslijnen uitgestippeld met betrekking tot de biodiversiteit van tropische regenwouden. Met betrekking tot het duurzaam beheer van tropische regenwouden hebben bijvoorbeeld de Internationale Organisatie van de Houthandel (ITTO) en het Centrum voor Internationaal Bosbouwkundig Onderzoek (CIFOR) initiatieven ontwikkeld om het behoud van de biodiversiteit als doelstelling op te nemen.

Biologische rijkdom: Biologische rijkdom wordt vaak gebruikt als een synoniem voor biodiversiteit. In zijn algemeenheid geldt dat biologische rijkdom hoog is in tropische laaglandgebieden en afneemt met toenemende lengtegraad en altitude (hoogte). Dit impliceert dat de tropische laagland gebieden zeer rijk zijn aan soorten. Dit behoeft echter niet per definitie waar te zijn en geldt zelfs niet voor alle tropische laaglandbossen. Diversiteit en rijkdom variëren per geografische regio en zijn afhankelijk van het abiotische milieu (niet-levende component), het biotische milieu (levende component) en van de evolutionaire, geologische en verstoringsgeschiedenis (Fig. 1). Als gevolg van al deze factoren is er een ruimtelijke variatie in de biodiversiteit en een variatie in de tijd.

Natuurbehoud: Aangezien het begrip biodiversiteit ontstaan is vanuit een wereldwijde bezorgdheid over het verdwijnen van soorten en ecosystemen, is het begrip ook nauw verwant met natuurbescherming. Bij het beschermen van natuur gaat het vaak over de waarde van die natuur. In dit verband wordt biodiversiteit vaak gezien als een waarde indicator met, naast een intrinsieke waarde, ook een economische 'optie' waarde. Deze optie waarde is dan een maat voor de potentiële waarde van biodiversiteit, een waarde die ontstaat vanuit mogelijke gebruiksmogelijkheden in de toekomst. Eigenlijk behelst een dergelijke optiewaarde een mogelijke nieuwe bedreiging voor de biologische diversiteit.

Tropenbos: De Nederlandse Stichting Tropenbos heeft tot doel om bij te dragen aan het behoud en duurzaam beheer van tropische regenwouden overal ter wereld door instrumenten en beleidskaders te ontwikkelen en aan te dragen voor beleidsmakers en bosbouwers. De Stichting heeft als tweede doel om in de landen waar zij actief is de lokale onderzoeksinstituten te betrekken bij het onderzoek en deze te versterken. Onder haar supervisie worden er onderzoeksprogramma's uitgevoerd in Kameroen, Colombia, Ivoorkust, Indonesië en Guyana. Het programma in Guyana is van start gegaan in 1989 en is een gezamenlijk project van de regeringen van Guyana en Nederland. Het Tropenbos-Guyana programma heeft specifiek tot doel om beleid ten aanzien van duurzaam bosgebruik en bosbeheer te ontwikkelen. Met betrekking tot het gebruik van het bos door de plaatselijke bevolking, met een nadruk op niet-hout bosproducten is er recentelijk een nieuw onderzoeksprogramma van start gegaan in het Noordwest district van Guyana. Niet-hout bosproducten is een term voor alle andere producten die uit het bos gebruikt worden, zoals voedsel, vruchten, medicinale planten en planten(delen) voor klein huishoudelijk gebruik.

Het grootste deel van het Tropenbos-Guyana onderzoek speelt zich echter af in een houtkapconcessie in het centrale deel van Guyana, en het onderzoek daar richt zich vooral op de duurzame extractie van tropisch hardhout. De houtkapconcessie wordt momenteel beheerd door Demerara Timbers Ltd. (DTL). Op de concessie bestudeert een team van Guyanese, Nederlandse en Engelse onderzoekers vanuit verschillende disciplines de abiotische en biotische onderdelen van het tropische ecosysteem en de samenhang tussen de verschillende onderdelen. De resultaten van de eerste fase van het onderzoeksprogramma (1989-1993) en de daaruit voortvloeiende aanbevelingen voor duurzaam bosbeheer zijn te vinden in Ter Steege et al. (1996).

Biodiversiteit

Het ontstaan en de handhaving van biodiversiteit

Tot nu is biodiversiteit in zijn algemeenheid behandeld. Dit proefschrift behandelt alleen botanische diversiteit en vanaf nu wordt er met biodiversiteit de soortensamenstelling en rijkdom van planten in het tropisch regenwoud bedoeld. Om te beginnen is daar de vraag waarmee de biodiversiteit van een gebied samenhangt. Een groot aantal factoren is daarop van invloed. In een schema (Fig. 1.1) is getracht aan te geven dat het aantal soorten dat in een gebied voorkomt afhankelijk is van zowel historische processen (het ontstaan van soorten) als van hedendaagse ecologische processen (de handhaving van soorten). Onder historische processen vallen de geologische geschiedenis (bv. plaattectoniek en ijstijden), de evolutionaire geschiedenis (processen van soortsvorming en uitsterving) en de regionale patronen van klimaat en regenval. Deze lange-termijn factoren worden van belang beschouwd voor de soortenrijkdom in een regio. Van de hedendaagse ecologische processen worden microklimaatsfactoren en verstoring het meest genoemd als van belang om verschillen op kleinere schaal (zowel in tijd als ruimte) te kunnen verklaren. Eigenlijk kunnen de meeste processen gezien worden als verstoringen en spelen deze zich af op vier verschillende niveaus. De geologische en evolutionaire geschiedenis vallen dan onder verstoringen op een zeer grote tijdsschaal en met een zeer grote reikwijdte (niveau A in Fig. 1.1). In een categorie verstoringen met een iets kleinere reikwijdte zouden dan landverschuivingen, vulkaanuitbarstingen, aardbevingen, orkanen, branden, droogte en grootschalige meanderende riviersystemen geplaatst kunnen worden. Het openen van het kronendak (hoogste niveau in het bos, niveau van de boomkronen) door het geheel of gedeeltelijk omvallen van een boom zou dan de kleinst mogelijke verstoringsgraad vormen. Verschillende abiotische factoren, zoals: bodemvruchtbaarheid, waterhuishouding, temperatuur en licht (niveau B in Fig. 1.1), worden ook gezien als essentiële factoren in het bepalen van het niveau van de soortenrijkdom van planten. Deze abiotische factoren kunnen op hun beurt weer de mate van verstoring (niveau C in Fig. 1.1) in het ecosysteem beïnvloeden. Als voorbeeld: bomen die voorkomen op ondiepe, nattere bodems zullen eerder omvallen dan bomen die op diepere en minder zompige bodems voorkomen. Ook is het voor te stellen dat in gebieden waar maar weinig water aanwezig is, een bosbrand die ontstaat door het inslaan van bliksem meer effect zal hebben dan in een veel nattere vegetatie. Het proces van het openen van het kronendak door het geheel of gedeeltelijk afsterven van bomen is een mooi voorbeeld van hoe biodiversiteit van een ecosysteem ook zelf invloed kan uitoefenen op de mate van verstoring in dat ecosysteem. Het biotische microklimaat en de zelfregulerende processen die optreden (bijv. concurrentie, voortplanting, groei en sterfte) worden in figuur 1.1 zeer summier aangegeven als niveau D. In figuur 2 is dit deel verder uitgewerkt.

Het belang van biodiversiteit

De optie waarde zoals hierboven besproken bij de inleiding geeft geen enkel aanknopingspunt in hoeverre biologische diversiteit of rijkdom belangrijk is voor het functioneren van tropische regenwouden. Wat is nu het belang van het aantal soorten dat deelneemt in een bepaald proces? Meer soorten zijn niet per definitie beter dan minder soorten om een bepaalde proces in het bos naar behoren te laten functioneren. Zo zijn er voorbeelden (zoals het behoud van bodemvruchtbaarheid) waarbij complexe natuurlijke ecosystemen (met veel soorten) even goed functioneren als een monocultuur (met maar 1 soort). Een ander voorbeeld wordt gegeven in het onderzoek van Leo Brouwer (Tropenbos-Guyana), die vond dat er na de houtkap in een opening in het bos rond Mabura veel meer nutriënten uit de bodem wegspoelden. Dit gebeurde vooral in die gebieden waar de vegetatie verdwenen was. Bij het dichtgroeien van de vegetatie nam deze uitspoeling weer af. Hierbij maakte het dan niet uit of de dichtgroeiende vegetatie uit veel soorten bestond, of uit maar een paar soorten.

Maar toch, ook al kunnen een paar soorten hetzelfde effect hebben op een biologisch proces dan veel meer soorten, het aantal soorten heeft een effect op de mate waarop de vegetatie kan reageren op verandering. Iedere soort heeft nl. andere basis behoeften ten aanzien van de omgeving, heeft een eigen wijze van verspreiding, een bepaalde groeisnelheid, en eigen maximale levensduur of oorzaken van sterfte. Bij een verandering in de omgeving zal iedere soort dan ook op een eigen wijze reageren. Dit houdt in dat soortensamenstelling een effect heeft op de mate van de biotische variabiliteit en daarbij (inderect) op processen in het ecosysteem. Soortensamenstelling kan ook een direct effect hebben op processen in het ecosysteem, zoals bijvoorbeeld de mate van vraat door herbivoren (planteneters) en de mate waarin strooisel gevormd wordt en wordt gedecomposteerd. Soortsdiversiteit voegt daarom robuustheid toe aan een ecosysteem, m.a.w. een ecosysteem met veel soorten is beter bestand tegen veranderende omstandigheden en kan zich daar beter aan aanpassen. Vrij recente experimenten laten ook zien dat de opname van kooldioxide en de productiviteit van planten (uitgedrukt in geproduceerde biomassa) verminderde met afnemende soortsdiversiteit. Dit suggereert dat een hoge soortsdiversiteit het functioneren van tropische bossen verbetert.

De relatie tussen biodiversiteit en houtkap

Houtkap heeft zowel een directe als een indirecte invloed op de biodiversiteit (zie figuur 1.3). Bij de extractie van bomen voor de houtoogst wordt het kronendak geopend en ontstaan er sleeppaden, daar waar zware machines de bomen het bos uit trekken. Vooral dit laatste leidt tot verdichting van de bodem, het omwoelen van de grond daar waar wortelsystemen uit de grond worden gerukt. Ook ontstaat er schade aan de overblijvende vegetatie. De diversiteit van de planten wordt direct beinvloed door het oogsten van bomen doordat sommige individuen worden verwijderd, sommige individuen worden bevoordeeld (bv. doordat ze meer licht krijgen waardoor ze sneller gaan groeien) en weer andere worden benadeeld (bv. doordat ze beschadigd zijn of doordat er voor hen teveel licht komt). Dit kan leiden tot een invloed op de soortensamenstelling van het bos.

Er kan ook sprake zijn van een indirecte invloed van de houtoogst op de biodiversiteit. Na de oogstcampagne is het bos toegankelijker voor jagers en andere 'verzamelaars' door de aanleg van wegen en paden. Dit kan leiden tot een verhoging van het aantal 'geoogste' soorten. Normalerwijs gesproken dragen planteneters (zoals agouti, peccari en tapir) bij aan de verspreiding van de planten, dan wel kunnen ze er door selectief te eten voor zorgen dat bepaalde planten niet te kans krijgen om te gaan domineren. Op hun beurt worden deze planteneters weer gegeten door roofdieren (zoals de jaguar). Wanneer de populatie van de planteneters, of hun predatoren, heel erg veranderd tengevolge van de druk van jagers, kan dit gevolgen hebben voor de plantengemeenschap. Ook het vangen en verzamelen van bijv. papegaaien (belangrijk voor de verspreiding van zaden) of niet-hout bosproducten kan bijdragen aan een verandering van de plantengemeenschap. Een ander indirect effect van de houtkap kan optreden als stukken bos onbereikbaar worden voor specifieke bestuivers of verspreiders. Ook de directe verandering van het abiotische milieu, zoals beschikbare hoeveelheid licht en nutriënten voor planten, kan leiden tot een veranderende concurrentieverhouding tussen planten, wat weer kan leiden tot een beïnvloeding van de biodiversiteit.

De selectieve houtkap kan dus een serie veranderingen teweeg brengen die ieder voor zich, of gezamenlijk, kunnen leiden tot een verandering van de biodiversiteit. In dit proefschrift wordt gekeken naar het resultaat van die verandering en wordt gepoogd een realtie te leggen tussen de mate van verstoring en de mate van verandering. Welke onderliggende factoren nu precies hebben bijgedragen aan de verandering is echter niet altijd bekend.

Een poging om de meest relevante aspecten van biodiversiteit in relatie tot de commerciële houtkap te begrijpen vereist om te beginnen een beeld van de soortensamenstelling van natuurlijk bos. Een tweede stap is het bestuderen van de gevolgen van houtoogst. Het is van belang om groepen van soorten te identificeren die op eenzelfde wijze reageren op de verstoring of díe soorten die een eventuele sleutelpositie innemen in het functioneren van het (bos)ecosysteem. Uiteindelijk zal er dan bekeken moeten worden of het mogelijk is om de reacties van de plantengemeenschap op de houtkap te voorspellen of dat de processen die door de houtkap in gang worden gezet zich niet houden aan wetmatigheden. In de onderhavige biodiversiteitsstudie wordt met duurzame houtkap bedoeld de duurzame extractie van commercieel interessante tropische hardhoutsoorten, waarbij er tegelijkertijd sprake is van handhaving van de karakteristieke soortensamenstelling van het bos.

Dus waar gaat dit proefschrift nu over

In het tropisch regenwoud van Guyana (Zuid-Amerika) is onderzoek gedaan naar de effecten van commerciële houtkap op de biodiversiteit (soortenrijkdom en -samenstelling van hogere planten) van het bos. Het onderzoek vond plaats in het kader van het Tropenbos-Guyana programma en werd gefinancierd door de Stichting Tropenbos, de Universiteit Utrecht en de Europese Unie. In Guyana wordt vooral Greenheart (Chlorocardium rodiei) geoogst, een tropische hardhoutsoort die veel wordt gebruikt in zeehavens. Het onderzoek heeft zich dan ook toegespitst op de verandering in botanische diversiteit van het bostype waar deze boom veel in voorkomt.

De doelstellingen van dit onderzoek waren:

1 het beschrijven van de biologische diversiteit van door Greenheart gedomineerd tropisch regenwoud en van de effecten van commerciële houtkap op deze biodiversiteit.
2 het bijdragen aan een duurzaam beheersplan voor door Greenheart gedomineerd tropisch regenwoud.
3 het toetsen van enkele theorieën over de soortenrijkdom van tropisch regenwoud.

Het onderzoek is op de volgende wijze uitgevoerd: Nabij Mabura Hill, de Tropenbos onderzoekslocatie in Centraal Guyana is in totaal van 24 ha bos de precieze soortensamenstelling opgenomen, zowel in ongestoord als in geëxploiteerd bos. De meeste studies over tropisch regenbos bestuderen alleen de bomen, in deze studie is echter zeer specifiek gekeken naar álle hogere planten: kruiden, struiken, palmen, lianen, epifyten en bomen. Van alle planten werd de grootte, diameter en precieze locatie in een proefvlak bepaald. Er is op vier verschillende locaties gewerkt. Om te beginnen is tijdens 1992 en 1993 van 15 ha ongestoord bos in Pibiri de soortensamenstelling en aantallen individuen gemeten van alle lianen, kruiden, struiken en kleine palmen. De bomen op deze lokatie zijn gemeten door Peter van der Hout, bosbouwer in het Tropenbos project. Pibiri is nl. ook de locatie van een ander Tropenbos onderzoek, een bosbouwkundige experiment om tijdens de houtoogst de kapschade te beperken en te komen tot een duurzaam beheersplan voor de exploitatie van Greenheart gedomineerd bos. De houtoogst op deze lokatie vond plaats in 1994. Begin 1996 zijn van twee ha opnieuw de soortenrijkdom en -samenstelling van alleen de lianen opgenomen. Tijdens de kapfase is de soorten-samenstelling en locatie van epifyten op verschillende gastheerboomsoorten gemeten.

Daarnaast is er gewerkt in drie andere geëxploiteerde gebieden: 2 km ten zuiden van Mabura Hill ('2km'), gekapt in 1985; 40 km ten Zuidoosten van Mabura Hill (Waraputa), gekapt in 1989-1990; net buiten het Tropenbos Reservaat (Reserve), gekapt in 1989. In elk van deze drie gebieden zijn 3 proefvlakken van 1 ha uitgezet, twee 1-ha proefvlakken in geëxploiteerd bos en één 1-ha proefvlak in natuurlijk bos. Het proefvlak in het natuurlijke bos werd geacht representatief te zijn voor de situatie van de geëxploiteerde proefvlakken vóór de kap. De geëxploiteerde gebieden zijn aan de studie toegevoegd om de invloed van de tijd na het kappen op de biodiversiteit te bestuderen. Door deze toevoeging werd er een extra niveau van geografische verscheidenheid geïntroduceerd. De houtoogst in de drie toegevoegde gebieden heeft plaats gevonden met gebruik van conventionele kapmethodes en met vergelijkbare kapintensiteiten. In de geëxploiteerde proefvlakken werd, naast de biodiversiteit, ook de locatie en grootte van kroonopeningen en sleeppaden gemeten. Met behulp van een Geografisch Informatie Systeem (GIS) werden van de proefvlakken computerrepresentaties gemaakt en werd achteraf geëvalueerd in welke boszone een plant zich bevond (bv. de afstand tot een sleeppad of tot de rand van een gat in het kronendak).

Hoofdstuk 1 bevat met een uitgebreide inleiding op het thema biodiversiteit en de mogelijke invloed van kappen op deze biodiversiteit. De verschillende betekenissen van het begrip 'biodiversiteit' worden behandeld en de doelstellingen en aanpak van deze studie worden uiteen gezet.

In **Hoofdstuk 2**wordt een blik geworpen op de regionale biodiversiteit, de totale flora van het Mabura Hill gebied. Om een idee te krijgen van de regionale en geografische achtergrond van het floragebied wordt om te beginnen de flora van het Mabura Hill gebied beschreven en vergeleken met de flora van andere locaties in Zuid-Amerika. Nadruk wordt gelegd op de vergelijking met de flora van andere locaties van het Guayana Schild en specifiek met andere locaties in de Guianas (Guyana, Suriname, Frans Guiana). In vergelijking met de andere locaties, en met name met twee Frans Guianese locaties, blijkt de flora van het Mabura Hill gebied vooral primaire regenwoudsoorten te bevatten, met een opmerkelijke schaarste aan ondergroei soorten (kruiden, struiken en kleine palmpjes) en epifyten. De locaties in de Guianas lijken echter wel het meeste op elkaar. Het tropische regenwoud in de omgeving van Mabura Hill is niet bijzonder rijk aan soorten en kent bijna geen geïntroduceerde of gebiedsvreemde soorten. Deze lage graad van geïntroduceerde soorten zou er op kunnen wijzen dat in het gebied weinig grootschalige verstoring heeft plaatsgevonden.

In dit hoofdstuk wordt geponeerd dat er een duidelijke verwantschap bestaat tussen de bosflora van West-Afrika en de flora van de Guianas, een sterkere verwantschap zelfs dan met andere delen van Zuid-Amerika. Deze stelling wordt onderbouwd met de herkomst van de meest belangrijkste botanische families in het gebied, het grote aandeel van de subfamilie Caesalpiniodeae in de Luguminosae én de hoge percentages van families zoals Connaraceae, Menispermaceae, Nyctaginaceae, Annonaceae en Apocynaceae. De lage verstoringsgraad van het gebied zou onderdeel kunnen zijn van een verklaring voor deze duidelijke verwantschap.

Hoofdstuk 3 bekijkt de diversiteitspatronen van een bepaalde groeivorm, de epifyten. De diversiteit van epifyten wordt op verschillende schalen geëvalueerd. Er wordt bekeken of epifyten specifiek zijn voor een bepaalde locatie, voor een bepaald bostype, of voor een bepaalde type gastheerboomsoort. De epifyten die in het Greenheart bos gevonden zijn werden daarvoor vergeleken met epifyten uit drie andere bostypen, zowel in Guyana als in Frans Guiana. Het bleek dat elk bostype zijn eigen specifieke soortensamenstelling had met het Greenheart bos als het bostype met het laagste aantal soorten. In elk van de vier bostypen waren het echter steeds dezelfde epifytensoorten die het meest voorkwamen. Die soorten die in minimaal drie van de vier bostypen voorkwamen vertoonden eenzelfde verdeling over een boom in elk bostype. De rijkste epifytische soortensamenstelling werd gevonden in de boomkronen van de verschillende bostypen en in de boomkronen leken de bostypen ook het meest op elkaar. In elk bostype konden vier verschillende epifytische leefmilieus worden onderscheiden, elk met een eigen groep epifytische soorten. De geografische afstand tussen verschillende bostypen speelde geen enkele rol in onderlinge verschillen in soortenrijkdom. De rijkdom en het voorkomen van epifyten lijkt vooral bepaald door het milieu. Vooral verschillen in vochtbeschikbaarheid en gastheerspecifieke factoren, zoals geschiktheid van de bast en boomarchitectuur, spelen hierin een belangrijke rol.

In **hoofdstuk 4** wordt er specifiek gekeken naar de diversiteit van Greenheart bos. Om te beginnen wordt er een beschrijving gegeven van de soortenrijkdom en -samenstelling van ongestoord Greenheart bos. De totale soortenrijkdom van Greenheart bos was, net als die voor de totale flora en voor de epifyten uit dit bostype, relatief laag. Verschillen in soortensamenstelling tussen en binnen verschillende ongestoorde proefvlakken kon voor de lianen (en in mindere mate voor de bomen) verklaard worden vanuit geografische (onderlinge) afstand. Voor ondergroei soorten werd dit verband niet aangetroffen.

Ieder groeivorm had eigen specifieke reacties op de houtkap. Het aantal soorten en individuen van lianen en bomen namen toe, terwijl er geen duidelijk verschil gevonden werd voor de groep van ondergroeisoorten (kruiden, struiken en kleine palmen). De gevonden reactiepatronen konden verschillen per observatieschaal (bv. een hele hectare of kapschade zones). Vooral de lianen bleken sterke correlaties te vertonen met verschillende vormen van kapschade. Ontstane gaten in het kronendak, sleeppaden, stukken bos met verhoogd lichtniveau ten gevolge van de kroonopeningen en 'ongestoord' bos hadden elk een eigen karakteristieke soortensamenstelling. Hoe groter het gebied dat verstoord was ten gevolge van de houtkap, des te meer nieuwe soorten er gevonden werden na de kap (vooral lianen en bomen). Ondergroeisoorten waren vrij eenvormig in soortensamenstelling, zowel in geëxploiteerd als in ongestoord bos, met een sterke dominantie van een paar (palmen) soorten. De soortensamenstelling in zowel kleine en grotere kroonopeningen (daar waar vooral de verjonging van het bos plaats vindt) bleek vooral afhankelijk van de soortensamenstelling van het omringende bos. Zes en tien jaar na de houtoogst waren de gevolgen nog steeds zichtbaar in een veranderde soortensamenstelling en verhoogd aantallen individuen. Er zijn geen indicaties gevonden dat deze toename in biodiversiteit weer afneemt op een tijdschaal van 10 jaar.

In **hoofdstuk 5** zijn de reacties van lianen op de houtoogst bekeken. Aangezien voor deze groeivorm de precieze soortensamenstelling van twee 1-ha proefvlakken van zowel voor als na de kap beschikbaar was, kon een meer gedetailleerde evaluatie plaats vinden. Vooral de ruimtelijke patronen van die individuen die zich na de kap gevestigd hadden, konden nu vergeleken worden met het voorkomen van verschillende kapschade zones. Duidelijke veranderingen in soortensamenstelling van lianen en hoeveelheid individuen konden worden aangetoond. Alhoewel er (in hoofdstuk 4) geen verschil werd gevonden tussen het aantal zeldzame soorten in ongestoord dan wel geëxploiteerd Greenheart bos, werd er voor de lianen wel een verschil waargenomen. Na de kap was het aandeel van de meer algemene lianensoorten veel groter dan die van de zeldzamere lianensoorten. Ongestoord Greenheart bos had een duidelijk andere soortensamenstelling met lianen met een voorkeur voor, of tolerantie ten opzichte van, lagere lichtniveaus. In de geëxploiteerde proefvlakken werden soorten dominant die een voorkeur vertoonden voor hogere lichtniveaus. Gebaseerd op het gedrag van individuele soorten werden een viertal functionele lianengroepen onderscheiden, elk met een eigen reactie op de houtoogst. Sommige lianensoorten verminderden in aantal individuen of verdwenen compleet, andere soorten vertoonden geen duidelijke reactie en weer andere soorten namen sterk toe in aantal individuen na de kap. De soorten die geen duidelijke reactie ten opzichte van de houtkap vertoonden konden vervolgens verdeeld worden in een groep van soorten met een duidelijke voorkeur voor hogere lichtniveaus en een groep van soorten met een duidelijke voorkeur voor lagere lichtniveaus. De verandering in lianen diversiteit van de twee proefvlakken die zowel voor als na de kap waren opgenomen, maakte duidelijk dat vooral in het deel van het bos met een verhoogd lichtniveau, maar zonder bodemverstoring, de grootste aanwas in soorten en individuen gevonden werd. Dit betrof die gebieden in het bos juist buiten de projectie van kroonopeningen op de bosbodem. Zes en tien jaar na de houtoogst werd deze aanwas in soorten en individuen niet langer aangetroffen in deze gebieden. Dit suggereert dat vooral in deze gebieden het aantal soorten en individuen na verloop van tijd terugloopt. De nieuw gevestigde soorten werden nu echter ook in andere gebieden aangetroffen. Deze veranderingen treden op op een kleiner spatiale schaal dan zoals geëvalueerd in hoofdstuk 4, en zijn niet zichtbaar op de schaal van 1-ha.

Hoofdstuk 6 is gewijdt aan het derde deel van de doelstelling van dit onderzoek. Verschillende gangbare theorieën over het ontstaan en de instant houding van soortsdiversiteit in de tropen worden getoetst aan de gevonden diversiteitspatronen van zowel de verschillende groeivormen in Greenheart bos, de epifyten en van de totale flora van het Mabura Hill gebied. Er bestaan veel verschillende theorieën over hoe het toch komt dat er in de tropen zoveel soorten voorkomen. Vaak wordt er binnen een theorie niet gedefinieerd op welke tijds- of ruimteschaal deze toepasbaar is, iets wat het toetsen van een dergelijke theorie niet ten goede komt. In dit hoofdstuk wordt dan ook getracht precies aan te geven op welke schaal er ondersteuning (of niet) voor de betreffende theorie gevonden werd. De verklaring voor hedendaagse niveaus van soortsdiversiteit wordt in het algemeen gezocht in een combinatie van historische en hedendaagse ecologische processen (Fig 1.1).

Van de beschikbare theorieën op evolutionaire tijdsschaal zou de 'stabiliteit-over-tijd' theorie een verklaring kunnen vormen voor het hedendaagse niveau van soortsdiversiteit van het tropisch regenwoud van de Guianas. Bij deze theorie wordt er van uit gegaan dat het feit dat tropische regenwouden al zeer lang bestaan en een stabiel klimaat bezitten geleid heeft tot een hoog niveau van soortsdiversiteit Op de ecologische tijdschaal kunnen de beschikbare theorieën verdeeld worden in twee belangrijke groepen: equilibriumtheorieën en niet-equilibriumtheorieën.

Equilibriumtheorieën gaan er vanuit dat bij een bepaalde combinatie van hulpbronnen sommige soorten het beter doen dan andere (harder groeien of beter overleven). Om een hoog aantal soorten te handhaven is het nodig dat de omgeving uit heel veel verschillende combinaties van hulpbronnen bestaat, of dat regelmatig terugkerende verstoringen er voor zorgen dat het bos steeds fragmenteert in een mozaïek van stukken bos in verschillende ontwikkelingsstadia. Dit in tegenstelling tot de niet-equilibriumtheorieën waar er vanuit gegaan wordt dat soorten eigenlijk helemaal niet van elkaar verschillen in hun vermogen om te reageren op de beschikbaarheid van hulpbronnen en dat de kans dat een van de soorten het 'beter doet' dan een andere erg klein is. De groep van soorten die in een bepaald gebied voorkomt is dan gewoon een kwestie van kans, een willekeurige trekking uit alle soorten die geschikt zijn om in dat gebied voor te komen.

De gegevens zoals beschikbaar uit het huidige onderzoek geven aan dat het onwaarschijnlijk is dat één en dezelfde theorie toepasbaar is op de verschillende groeivormen of soortengroepen, aangezien de gevonden patronen sterk verschillen per groeivorm. Geografische afstand is een belangrijke factor in het verklaren van verschillen in soortensamenstelling van lianen en in mindere mate van bomen. Voor epifyten en ondergroei soorten is dit echter niet het geval. De aanwezigheid van epifyten lijkt vooral bepaald door omgevingsfactoren. Zowel de veronderstelling dat de soortensamenstelling vooral bepaald zou worden door het mozaïek van stukken bos in verschillende ontwikkelingsstadia als de veronderstelling dat de soortensamenstelling een willekeurige trekking uit een grotere verzameling van soorten zou zijn, worden niet ondersteund door het duidelijke geografische patroon zoals gevonden bij de lianen.

Alhoewel er voor een paar lianensoorten wel een duidelijke voorkeur voor een specifieke ontwikkelingsfase van het mozaïek is vastgesteld, was de soortensamenstelling van de lianen toch vooral lokaal bepaald. D.w.z. dat iedere lokatie een eigen, specifieke, combinatie van soorten en aantallen individuen had. Bij verstoring en het ontstaan van kroonopeningen was het deze lokale combinatie die bepaalde welke soorten er tot wasdom zouden kunnen komen. Er werd maar een klein aantal soorten gevonden wat zich altijd vestigde na de kap, en deze maakte dan ook maar een klein deel uit van de totale soortensamenstelling.

De gegevens uit dit onderzoek waren niet geschikt om te toetsen in hoeverre interacties tussen individuen (concurrentie) belangrijk zijn voor het instandhouden van een hoog niveau van soortsdiversiteit in de ruimte. Dit mechanisme van interacties tussen individuen lijkt echter het enige mechanisme waarin duidelijk zowel de ruimte die individuen innemen en waar ze deel van uitmaken én het vermogen van individuen om van diezelfde ruimte de structuur en omgevingsfactoren te beïnvloeden een rol speelt. Toekomstig onderzoek (zowel in het veld als theoretisch) zal moeten uitmaken of deze interacties een bevredigende verklaring kunnen vormen voor zowel de verschillende reacties van planten op de houtoogst als voor de relatieve onbelangrijkheid van het mozaïek. Wat er vooral nodig is zijn gegevens op het gebied van gezamenlijke kenmerken van soorten in eenzelfde functionele groep, zoals bijvoorbeeld plantfysiologische reacties op de hulpbronnen licht en water, groeisnelheden van planten en de wijze van verspreiden. Slechts van een klein deel van de soorten die in Greenheart bos voorkomen zijn deze gegevens bekend en dat betreft dan bijna altijd boomsoorten. De eenzijdige aandacht voor bomen in het tropisch ecologische onderzoek gaat voorbij aan het feit dat concurrentie op het niveau van de bosbodem bepaald welke individuen de beschikbare ruimte zullen innemen. Deze concurrentie vindt plaats tussen alle individuen die zich op de bosbodem bevinden. Het wordt hoog tijd dat er meer aandacht komt voor de niet-bomen.

Indien er in de toekomst meer gegevens beschikbaar zijn over de precieze soortensamenstelling, op precies dezelfde plek, van voor en na de houtoogst, kunnen op een meer gedetailleerde wijze theorieën rond de instandhouding van soorten in het tropische regenwoud getoetst worden.

Ter invulling van de tweede doelstelling van dit onderzoek zijn er naar aanleiding van de resultaten van het onderzoek een negental aanbevelingen voor duurzaam bosbeheer geformuleerd (**Hoofdstuk** 7). Een aantal van deze aanbevelingen waren al eerder geformuleerd op grond van onderzoek vanuit andere disciplines of op basis van gezond verstand, en deze zijn nu verder onderbouwd met gegevens over de mogelijke verandering in soortensamenstelling van het bos ten gevolge van houtoogst. Zo wordt de aanbeveling om zo min mogelijk bomen per oppervlakteeenheid te oogsten onderbouwd met de relatie tussen het aantal gekapte bomen en het aantal lianensoorten na de houtoogst. Hoe meer bomen er geoogst worden, hoe meer lianensoorten na de oogst aanwezig zijn. Het aantal lianensoorten na de houtoogst vertoonde ook een sterke relatie met de verandering in de totale soortensamenstelling. Om het bos zoveel mogelijk in de oorspronkelijke samenstelling te laten is het dus aan te bevelen om zo min mogelijk bomen te oogsten. Een andere aanbeveling gaat over de invloed van de grootte van een kroonopening. Het bleek nl. dat de soortensamenstelling in kleinere kroonopeningen (kleiner dan 300 m^2) meer overeenkomst vertoonde met de soortensamenstelling van ongestoord bos dan de soortensamenstelling in grotere kroonopeningen (van meer dan 300 m^2). Het is daardoor aan te bevelen de ontstane gaten in het kronendak zo klein mogelijk te houden. Een andere belangrijke factor die het aantal soorten na de houtoogst beïnvloedde waren de sleeppaden. Het minimaliseren van het aantal en de grootte van sleeppaden is dan ook de volgende aanbeveling. Vooral het bosdeel waar zowel een opening in het kronendak als een sleeppad gevonden kon worden vertoonde de hoogste aantallen nieuwe soorten na de houtoogst, het rijden met zware machinerie in een kroonopening dient dan ook ze veel mogelijk te worden beperkt.

Verder zijn er specifieke aanbevelingen opgenomen t.a.v. lianen. Lianen worden nl. door veel bosbouwers gezien als 'hinderlijk'. Lianen kunnen bomen met elkaar verbinden en daardoor bij het omzagen van één boom meerdere bomen mee omtrekken waardoor het resulterende gat in het kronendak veel groter wordt. Het wegkappen van alle lianen enige tijd voor de houtkap is dan ook een gangbare bosbouwpraktijk. Uit het huidige onderzoek is gebleken dat maar een klein deel van de lianen in staat is meerdere bomen met elkaar te verbinden en een zodanige diameter te bereiken dat ze ook daadwerkelijk schade kunnen berokkenen. Het wegkappen van lianen hoeft enkel en alleen te geschieden bij deze groep van soorten. Een simpele identificatie sleutel voor deze soorten is nodig om te helpen bij het herkennen. Verder kunnen lianen in zeer verstoorde gebieden een 'lianen deken' vormen, een ondoordringbare kluwen van opgroeiende lianen. Opnieuw zijn het maar een paar soorten die in staat zijn een dergelijke deken te vormen. De vestiging van andere plantensoorten (en eventueel van commercieel 'gewenste' soorten) kan echter hinder ondervinden van een 'lianen deken'. Om dit proces tegen te gaan wordt aanbevolen om, alleen als het verschijnsel zich voordoet, beperkte bosbouwkundige maatregelingen te treffen. Aangezien het verschijnsel 'lianen deken' zich alleen voordoet in stukken bos met heel veel kapschade, zal een dergelijke deken niet ontstaan bij voldoende maatregelen om de kapschade terug te dringen.

References

Ackerman, J.D., A.M. Montalvo & A.M. Vera. 1989. Epiphyte host specificity of *Encyclia krugii*, a Puerto Rican endemic orchid. *Lindleyana* 4: 74-77.

Alder, D. & T.J. Synnott. 1992. Permanent sampling plot techniques for mixed tropical forest. *Tropical Forestry Paper* 25. Oxford Forestry Institute. University of Oxford. UK.

Appanah, & F.E. Putz. 1984. Climber abundance in virgin Dipterocarp forest and the effect of pre-felling climber cutting on logging damage. *Malaysian forester* 47: 335-342.

Axelrod, D.I & P.H. Raven. 1978. Late Cretaceous and Tertiary vegetation history of Africa. *In:* Werger, M.J.A. (ed.). *Biogeography and ecology of Southern Africa.* W. Junk. The Hague.

Balfour, D.A. & W.J. Bond. 1993. Factors limiting climber distribution and abundance in a southern African forest. *J. Ecol.* 81: 93-99.

Balée, W. & D.G. Campbell. 1990. Evidence for the successional status of liana forest (Xingu River Basin, Amazonian Brazil). *Biotropica* 22(1): 36-47.

Bennett, B.C. 1986. Patchiness, diversity and abundance relationships of vascular epiphytes. *Selbyana* 9: 70-75.

Benzing, D.H. 1981. Bark surfaces and the origin and maintenance of diversity among angiosperm epiphytes: An hypothesis. *Selbyana* 5: 248-255.

Benzing, D.H. 1989. Vascular epiphytism in America. *In:* Lieth, H. & M.J.A. Werger (eds.). *Tropical rain forest ecosystems. Biogeography and Ecology. Ecosystems of the World* 14A. Elsevier, Amsterdam, The Netherlands.

Benzing, D.H. 1990. *Vascular epiphytes.* Cambridge University Press, Cambridge.

Benzing, D.H. 1995. Vascular epiphytes. *In:* Lowman, E.D. & N.M. Nadkarni (eds.). *Forest canopies.* Academic Press, San Diego, USA.

Berry, P.E., O. Huber & B.K. Holst. 1995. Floristic analysis and phytogeography. *In:* Berry, P.E., B.K. Holst & K. Yatskievych (eds.). *Flora of the Venezuelan Guayana. Vol 1: Introduction.* Missouri Botanical Garden, St. Louis and Timber Press, Portland, Oregon.

Besag, J. & P. Clifford. 1989. Generalized Monte Carlo significance tests. *Biometrika* 76: 633-642.

Beven, S., E.F. Connor & K. Beven. 1984. Avian biogeography in the Amazon basin and the biological model of diversification. *J. Biogeogr.* 11: 383-399.

Boggan, J., V. Funk, C. Kelloff, M. Hoff, G. Cremers, & C. Feuillet. 1992. *Checklist of the plants of the Guianas.* Biological Diversity of the Guianas Program. Washington D.C.

Boggan, J., V. Funk, C. Kelloff, M. Hoff, G. Cremers, & C. Feuillet. 1997. *Checklist of the plants of the Guianas (Guyana, Surinam, French Guiana). 2nd Edition.* Biological Diversity of the Guianas Program. Washington D.C.

Boot, R.G.A. 1994. Growth and survival of tropical rain forest tree seedlings in forest understorey and canopy openings. Implications for management. *Tropenbos Documents 6.* Tropenbos, Wageningen. The Netherlands.

Borcard, D., P. Legendre & P. Drapeau. 1992. Partialling out between the spatial component of ecological variation. *Ecology* 73: 1045-1055.

Bordenhave, B. 1996. *Mesures de la diversite specifique des plantes vasculaires en forêt sempervirente de Guyane.* Unpublished Thesis, Muséum National d'Histoire Naturelle, Paris.
Brewer, A. & M. Williamson. 1994. A new relationship of rarefaction. *Biodiv. Conserv.* 3: 373-379
Brokaw, N.V.L. 1985a. Treefalls, regrowth and community structure in tropical forests. *In:* S.T.A. Pickett & P.S. White (eds.). *The ecology of natural disturbance and patch dynamics.* Academic Press. Orlando.
Brokaw, N.V.L. 1985b. Gap phase regeneration in a tropical forest. *Ecology* 66: 682-687.
Brokaw, N.V.L. 1987. Gap-phase regeneration of three pioneer tree species in a tropical forest. *J. Ecol.* 75: 9-19.
Brouwer, L.C. 1996. *Nutrient cycling in a tropical rain forest in Guyana.* Ph.D. thesis, University of Utrecht. Tropenbos Guyana Series Number 1, Georgetown, Guyana.
Brown, N. & M. Press. 1992. Logging rainforests the natural way? *New Scientist* 14 March 1992.
Bruenig, E.F. 1996. *Conservation and management of tropical rainforests. An integrated approach to sustainability.* CAB International, Wallingford, UK.
Bruijnzeel, L.A. 1990. *Hydrology of moist tropical forests and effects of conversion: a state of knowledge review.* UNESCO Humid Tropics Programme, Paris.
Bruijnzeel, L.A. & W.R.S. Critchley. 1994. *Environmental impacts of logging moist tropical forests.* IHP Humid Tropics Programme Series No. 7. UNESCO, Paris, France.
Burghouts, T.B.A., E.J.F. Campbell & P.J. Kolderman. 1994. Effects of tree species heterogeneity on leaf fall in primary and logged Dipterocarp forest in the Ulu Segama Forest Reserve, Sabah, Malaysia. *J. Trop. Ecol.* 10: 1-26.
Cain, S.A., G.M. de O. Castro, J.M. Pires & N.T. da Silva. 1956. Application of some phytosociological techniques to Brazilian rain forest. *Amer. J. Bot.* 43: 911-941.
Campbell, D.G., D.C. Daly, G.T. Prance & U.N. Maciel. 1986. Quantitative ecological inventory of terra firme and varzea tropical forest on the Rio Xingu, Brazilian Amazon. *Brittonia* 38(4): 369-393.
Catling, P.M. & L.P. Lefkovich. 1989. Association of vascular epiphytes in a Guatemalan cloud forest. *Biotropica* 21: 35-40.
Catling, P.M., V.R. Brownwell & L.P. Lefkovich. 1986. Epiphyte orchids in a Belizean grapefruit orchard: distribution, colonization, and association. *Lindleyana* 1: 194-202.
Chazdon, R.L., R.W. Pearcy, D.W. Lee & N. Fletcher. 1996. Photosynthetic responses of tropical forest plants to contrasting light environments. *In:* Mulkey, S.S., R.L. Chazdon & A.P. Smith (eds.). *Tropical forest plant ecophysiology.* Chapman & Hall, New York.
Chazdon, R.L., R.K. Colwell, J.S. Denslow & M.R. Guariguata (in press). Statistical estimation of species richness of woody regeneration in primary and secondary rain forests in NE Costa Rica. *In: Proceedings of the SI/MAB International Symposium on Measuring and Monitoring Forest biological diversity,* 1995.
Clark, D.B. & D.A. Clark. 1990. Distribution and effect on tree growth of lianas and woody hemiepiphytes in a Costa Rican tropical wet forest. *J. Trop. Ecol.* 6: 321-331.
Clinebell, R.R., O.L. Phillips, A.H. Gentry, N. Stark & H. Zuuring. 1995. Prediction of neotropical tree and liana species richness from soil and climatic data. *Biodiv. Conserv.* 4: 56-90.
Cody, M.L. 1986. Diversity, rarity, and conservation in mediterranean-climate regions. *In:* Soulé, .E. (ed.). *Conservation biology: the science of scarcity and diversity.* Sinauer, Sunderland, Massachusetts.
Colinvaux, P. 1993. Pleistocene biogeography and diversity in tropical forests of South America. *In:* Goldblatt, P. (ed.). *Biological relationships between Africa and South America.* Yale University Press, New Haven & London.
Comiskey, J.A., G. Aymard & F. Dallmeier. 1994. Structure and composition of lowland mixed forest in the Kwakwani region of Guyana. *Biollania* 10: 13-28.
Condit, R., S.P. Hubell & R.B. Foster. 1994. Density dependance in two understorey tree species in a neotropical forest. *Ecology* 75: 209-233.
Condit, R., S.P. Hubbell, J.V. LaFrankie, R. Sukumar, N. Manokaran, R.B. Foster & P.S. Ashton. 1996. Species-area and species-individual relationships for tropical trees: a comparison of three 50-ha plots. *J. Ecol.* 84: 549-562.

Connell, J.H. 1971. On the role of natural enemies in preventing competitive exclusion in some marine animals and in rainforest trees. *In:* den Boer, P.J. & G.R. Gradwell (eds.). *Dynamics of populations.* Centre for Agricultural Publishing, Wageningen, The Netherlands.

Connell, J.H. 1978. Diversity in tropical rain forests and coral reefs. *Science* 199: 1302-1310.

Connell, J.H. & M.D. Lowman. 1989. Low-diversity tropical rain forests: some possible mechanisms for their existence. *American Naturalist* 134: 88-119.

Cowan, R.S. & J.C. Lindeman. 1989. *Ceasalpiniaceae p.p. In:* Görts van Rijn, A.R.A. (ed.). Flora of the Guianas. Series A: Phanerogams: 88. Koeltz Scientific Books, Koenigstein, FRG.

Coxson, D.S. 1991. Nutrient release from epiphytic bryophytes in tropical montane rain forest Guadeloupe. *Canad. J. Bot.* 69: 2122-2129.

Coxson. D.S., D.D. McIntyre & H.J. Vogel. 1992. Pulse release of sugars and polyols from canopy bryophytes in Tropical montane rain forest Guadeloupe, French West Indies. *Biotropica* 24(2a): 121-133.

Cremers, G., C. Feuillet, C.A. Gracie, J.J. de Granville, M. Hoff & S.A. Mori. 1993. Liste de Phanérogames et Ptéridophytes de la région de Saül, Guyane Française. *Gestion de l'écosysteme forestier et aménagement de l'espace régional.* Actes du IIéme Congrès Régional de L'environment . Sepanguy, Cayenne, Guyane Française.

Cronquist, A. 1981. *An integrated system of classification of flowering plants.* Colombia University Press, New York.

Connor, E.F. & E.D. McCoy. 1979. The statistics and biology of the species-area relationship. *Amer. Naturalist* 113: 791-833.

Davis, T.A.W. & P.W. Richards. 1932. The vegetation of Moraballi creek, British Guiana. An ecological study of a limited area of tropical rain forest I. *J. Ecol.* 21: 350-384.

Davis, T.A.W. & P.W. Richards. 1934. The vegetation of Moraballi creek, British Guiana. An ecological study of a limited area of tropical rain forest II. *J. Ecol.* 22: 106-115.

De Granville, J.J. 1988. Phytogeographical characteristics of the Guianan forests. *Taxon* 37: 578 -594.

De Koning, J. 1983. *La Forêt du Banco.* PhD thesis. Meded. Landbouwhogeschool Wageningen 83-1.

DeMenocal, P.B. 1995. Plio-Pleistocene African climate. *Science* 270: 53-58.

Denslow, J.S. 1980. Gap partitioning among tropical rainforest trees. *Biotropica* 12(Suppl.): 47-55.

Denslow, J.S. 1995. Disturbance and diversity in tropical rain forests: The density effect. *Ecol. Appl.* 5(4): 962-968.

Denslow, J.S. 1996. Functional group diversity and responses to disturbance. *In:* Orians, G.H., R. Dirzo & J.H. Cushman (eds.). *Biodiversity and ecosystem processes in tropical forests.* Ecological Studies 122: 127-151.

Deutschman, D.H., S.A. Levin, C. Devine & L.A. Buttel. 1997. Scaling from trees to forests. Am. Ass. Advanc. Science. Via Internet: http://www.sciencemag.org/feature/data/ deutschmann.

Dressler, R.L. 1981. *The orchids. Natural history and classification.* Harvard Univ. Press, Cambridge.

Dirzo, R., C.C. Horvitz, H. Quevedo & M.A. Lopez. 1992. The effects of gap size and age on the understorey herb community of a tropical Mexican rain forest. *J. Ecol.* 80: 809-822.

Duivenvoorden, J.F. & J.M. Lips. 1995. *A land-ecological study of soils, vegetation and plant diversity in Colombian Amazonia.* Tropenbos Series 12. The Tropenbos Foundation, Wageningen, The Netherlands.

EEI. Essence Economics Institute 1997. "Rio Cluster" of United Nations proceedings. *Internet site,* Information Habitat. http://www.igc.apc.org/habitat/un-proc.

Ek, R.C. 1990. *Index of Guyana plant collectors. In:* Görts van Rijn, A.R.A. (ed.). Flora of the Guianas. Supplementary series 1. Koeltz Scientific Books, Koenigstein, FRG.

Ek, R.C., C.A. Van der Kooij & J.A.C. Van Dam. *submitted.* The effect of logging on the liana vegetation in Greenheart dominated tropical rain forest near Mabura Hill, Guyana.

Ek, R.C., H. Ter Steege & J.C. Biesmeijer. Submitted. Vertical distribution and associations of vascular epiphytes in four different forest types in the Guianas.

Eiten, G. 1992. How names are used for vegetation. *J. Veg. Sci.* 3: 417-424.

Emmons, L.H. & A.H. Gentry. 1983. Tropical forest structure and the distribution of gliding and prehensile-tailed vertebrates. *Amer. Naturalist* 121: 513-524.

Evans, C. & B.J. Meggers. 1960. *Archeological investigations in Bristish Guiana.* Bureau of Amerindian Ethnology Bulletin 177, Washington DC., USA.

Ewel, J.J. & S.W. Bigelow. 1996. Plant life-forms and tropical ecosystem functioning. *In:* Orians, G.H., R. Dirzo & J.H. Cushman (eds.). *Biodiversity and ecosystem processes in tropical forests.* Ecological Studies 122: 101-126.

Ewel, J.J., M.J. Mazzarino & C.W. Berish. 1991. Tropical soil fertility changes under monocultures and successional communities of different structure. *Ecol. Appl.* 1: 289-302.

Fanshawe, D.B. 1952. *The vegetation of British Guyana. A preliminary review.* Institute Papers: 29. Imperial Forestry Institute, Oxford, UK.

Fisher, J.B. & F.W. Ewers. 1991. Structural responses to stem injury in vines. *In:* Putz, F.E. & H.A. Mooney (eds). *The biology of vines.* Cambridge University Press, Cambridge, UK.

Fjeldså, J. & J.C. Lovett. 1997. Geographical patterns of old and young species in African forest biota: the significance of specific montane areas as evolutionary centres. *Biodiv. Conserv.* 6: 325-346.

Forero, E. 1983. Connaraceae. Flora Neotropica Monograph No. 36. The New York Botanical Garden. New York.

Foster, R.B. 1990.The floristic composition of the Rio Manu Floodplain Forest. *In:* Gentry, A.H. (ed.). *Four Neotropical Rain Forests.* Yale University Press, New Haven & London.

Fox, J.E.D. 1968. Logging damage and the influence of climber cutting prior to logging in the lowland Dipterocarp forest of Sabah. *Malaysian forester* 31(4): 326-347.

Frei, S.J.K. 1973. Orchid ecology in a cloud forest in the mountains of Oaxaca, Mexico. Amer. *Orchid Soc. Bull.* 42: 307-314.

Freiberg, M. 1996. Spatial distribution of vascular epiphytes on three emergent canopy trees in French Guiana. *Biotropica* 28(3): 345-355.

Galetti, M. & F. Pedroni. 1994. Seasonal diet of capuchin monkey (*Cebus appela*) in a forest in South-east Brazil. *J. Trop. Ecol.* 10: 27-39.

Gaston, K.J. 1996. Species richness: measure and measurement. *In:* K.J. Gaston (ed.). *Biodiversity, a biology of numbers and difference.* Blackwell Science. Oxford. UK.

Gentry, A.H. 1988. Changes in plant community diversity and floristic composition on environmental and geographical gradients. *Ann. Missouri Bot. Gard.* 75: 1-34.

Gentry, A.H. 1990. Floristic similarities and differences between Southern Central America and Upper and Central Amazonia. *In:* Gentry, A.H. (ed.). *Four Neotropical Rain Forests.* Yale University Press, New Haven & London.

Gentry, A.H. 1991a. Reproductive strategies of lianas. *In:* Putz, F.A. & H.A. Mooney (eds.). *The biology of vines.* Cambridge University Press, Cambridge, UK.

Gentry, A.H. 1991b. The distribution and evolution of climbing plants. *In:* Putz, F.A. & H.A. Mooney (eds.). *The biology of vines.* Cambridge University Press, Cambridge, UK.

Gentry, A.H. 1993. Diversity and floristic composition of lowland tropical forest in Africa and South America. *In:* Goldblatt, P. (ed.). *Biological relationships between Africa and South America.* Yale University Press, New Haven, USA.

Gentry, A.H. & C.H. Dodson. 1987. Contribution of non-trees to species richness of a tropical rain forest. *Biotropica* 19 (2): 149-156.

Gentry, A.H. & C.H. Dodson. 1988. Diversity and biogeography of neotropical vascular epiphytes. *Ann. Missouri Bot. Gard.* 74: 205-233.

Goldblatt, P. 1993. *Biological relationships between Africa and South America.* Yale University Press, New Haven, USA.

Grubb, P.J. 1977. The maintenance of species-richness in plant communities: the importnace of the regeneration niche. *Biol. Rev. Biol. Proc. Cambridge Philos. Soc.* 52: 107-145.

Haffer, J. 1974. Avian speciation in tropical South America. *Publs. Nuttal. Ornithol. Club* 14: 1-390.

Haffer, J. 1982. General aspects of the refuge theory. *In:* Prance, G.T. (ed.). *Biological diversification in the tropics.* Columbia University Press. New York.

Haffer, J. 1987. Quarternary history of tropical America. *In:* Whitmore, T.C. & G.T. Prance (eds.). *Biogeography and quaternary history in tropical America.* Clarendon Press. London.
Haffer, J. 1997. Alternative models of vertebrate speciation in Amazonia: an overview. *Biodiv. Cons.* 6(3): 451-476.
Hammond, D.S. & V.K. Brown. 1991. *The ecological basis of recruitment and maintenance of timber tree species in the forests of Guyana.* Internal Group Report DSH3. Imperial College, Ascot, UK.
Hammond, D.S. & V.K. Brown. 1995. Seed size of woody plants in relation to disturbance, dispersal, soil type in wet Neotropical forests. *Ecology* 76 (8): 2544-2561.
Hammond, D.S. & V.K. Brown. In press. Disturbance, phenology and life-history characteristics: factors influencing frequency-dependent invertrebrate attack on tropical seeds and seedlings. *In:* Newbery, D.M., H.H.T. Prins & N.D. Brown (eds.). *Population and community dynamics in the tropics.* Blackwell Scientific Publications, Oxford.
Hartshorn, G.S. 1978. Tree falls and tropical forest dynamics. *In:* Tomlinson, P.B. & M.M. Zimmermann (eds.). *Tropical trees as living systems.* Cambridge University Press, Cambridge: 617-638.
Hartshorn, G.S. 1980. Neotropical forest dynamics. *Biotropica* 12 (Suppl.): 23-30.
Hartshorn, G.S. & B.E. Hammel. 1994. Vegetation types and floristic patterns. *In:* McDade, L.A., S.B. Kamaljit, H.A. Hespenheide & G.S. Hartshorn (eds.). *La Selva: ecology and natural history of a Neotropical rainforest.* University of Chicago Press, Chicago.
Hawthorne, W.D. 1993. *Forest regeneration after logging.* ODA Forestry Series No. 3.
Hegarty, E.E. & G. Caballé. 1991. Distribution and abundance of vines in forest communities. In Putz, F.E. & H.A. Mooney (eds.). *The biology of vines.* Cambridge University Press, Cambridge, UK.
Heywood, V.H. 1985. *Flowering plants of the world.* Equinox Ltd., Oxford, UK.
Hogeweg, P. & B. Hesper. 1972. *BIOPAT.* Program system for bioinformatic pattern analysis. Dept. of Theoretical Biology, Utrecht University, The Netherlands.
Hogeweg, P. & B. Hesper. 1981. Oligothetic characterization of clusters. *Pattern recognition* 14: 131-136.
Hogeweg, P. 1994. Multilevel evolution: replicators and the evolution of diversity. *Physica* D 75: 275-291.
Hoogmoed, M.S. & R. de Jong. 1992. *Tropisch regenwoud, schatkamer van diversiteit.* Natuur Historisch Museum, Leiden.
Hubbell, S.P. 1979. Tree dispersion, abundance, and diversity in a tropical dry forest. *Science* 203: 1299-1309.
Hubbel, S.P. & R.B. Foster. 1983. Diversity of canopy trees in a neotropical forest and implications for conservation. *In:* Sutton, S.L., T.C. Whitmore & A.C. Chadwick (eds.). *Tropical rain forest: Ecology & management.* Blackwell Scientific Publications, Oxford.
Hubbell, S.P. & R.B. Foster. 1986. Biology, change and history and the structure of tropical rain forest communities. *In:* Diamond, J. & T.J. Case (eds.). *Community ecology:* pp. 314-329. Harper & Row, New York.
Hubbell, S.P & R.B. Foster. 1990. Structure, dynamics and equilibrium status of old-growth forest on Barro Colorado Island. *In:* Gentry, A.H. (ed.). *Four Neotropical Rain Forests,* pp. 522-541. Yale University Press, New Haven & London.
Hubbell, S.P. & R.B. Foster. 1992. Short-term dynamics of a Neotropical forest: why ecological research matters to tropical conservation and management. *Oikos* 63: 48-61.
Huber, O. 1995. Vegetation. *In:* Berry, P.E., B.K. Holst & K. Yatskievych (eds.). *Flora of the Venezuelan Guayana. Vol. 1: Introduction.* Missouri Botanical Garden, St. Louis and Timber Press, Portland, Oregon.
Huston, M.A. 1979. A general hypothesis of species diversity. *Amer. Naturalist* 113: 81-101.
Huston, M.A. 1994. *Biological diversity.* The coexistence of species on changing landscapes. Cambridge University Press. Cambridge. UK.
Huston, M.A. & L.E. Gilbert. 1996. Consumer diversity and secondary production. *In:* Orians, G.H., R. Dirzo & J.H. Cushman (eds.). *Biodiversity and ecosystem processes in tropical forests.* Ecological Studies 122. Pp. 33-47. Springer Verlag. Berlin.
Huston, M.A. & T.M. Smith. 1987. Plant succession: Life history and competition. *Amer. Naturalist* 130: 168-198.

Jaccard, P. 1912. The distribution of the flora of the alpine zone. *New Phytologist* 11: 37-50.

Jans, L., L. Poorter, R.S.A.R. Van Rompaey & F. Bongers. 1993. Gaps and forest zones in tropical moist forest in Ivory-Coast. *Biotropica* 25(3): 258-269.

Janzen, D.H. 1970. Herbivores and the number of tree species in tropical forests. *Amer. Naturalist* 104: 501-528.

Jetten, V.G. 1994. *Modeling the effects on the water balance of a tropical rain forest.* A study in Guyana. PhD thesis, Utrecht University. Tropenbos series 6. The Tropenbos Foundation, Wageningen, The Netherlands.

Johansson, D. 1974. Ecology of vascular epiphytes in West African rain forest. *Acta Phytogeogr. Suec.* 59: 1-136.

Johns, A.D. 1992. Species conservation in managed tropical forests. *In:* Whitmore, T.C. & J.A. Sayer (eds.). *Tropical deforestation and species extinction.* Chapman & Hall, London, UK.

Johnson, S.P. 1993. *The earth summit: the United Nations Conference on Environment and Development (UNCED).* Graham and Trotman, London.

Johnston, M. & M. Gillman. 1995: Tree population studies in low-diversity forests, Guyana. I. Floristic composition and stand structure. *Biodiv. Conserv.* 4: 339-362.

Jongkind, C.C.H. & R.H.M.J. Lemmens. 1989. *The Connaraceae. A taxonomic study with special emphasis on Africa.* Agric. Univ. Wageningen Papers 89-6.

Jordan, C.F. 1985. *Nutrient cycling in tropical forest ecosystems.* Wiley, Chichester, UK.

Kalliola, R., M. Puhakka, J. Salo, H. Tuomisto & K. Ruokolainen. 1991. The dynamics, distribution and classification of swamp vegetation in Peruvian Amazonia. *Ann. Bot. Fenn.* 28: 225-239.

Kapos, V., E. Pallent, A. Bien & S. Freskos. 1990. Gap frequencies in lowland rain forest sites on contrasting soils in Amazonian Ecuador. *Biotropica* 22(3): 218-225.

Korning, J. & H. Balslev. 1994. Growth and mortality of trees in Amazonian tropical rain forest in Ecuador. *J. Veg. Sci.* 4: 77-86.

Lawton, R.O. & F.E. Putz. 1988. Natural disturbances and gap-phase regeneration in a wind-exposed tropical cloud forest. *Ecology* 69(3): 764-777.

Lawton, J.H. & V.K. Brown. 1993. Redundancy in ecosystems. *In:* E.D. Schulze & H.A. Mooney (eds.) *Biodiversity and ecosystem function.* Ecological Studies 99. Pp. 255-270. Springer Verlag, Berlin.

Lescure, J.P., H. Puig, B. Riera & D. Sabatier. 1990. Une forêt primaire du Guyane française: données botaniques. *In:* Sallailh, J.M. (ed.). *Mise en valeur de l'écosystème forestier guyanais.* Opération ECEREX. INRA, Paris & CTFT, Nogent-sur-Marne.

Lewis, G.P. 1987. *Legumes of Bahia.* Royal Botanic Gardens, Kew, UK.

Lieberman, M., D. Lieberman & R. Peralta. 1989. Forests are not just swiss cheese: canopy stereo geometry of non-gaps in tropical forests. *Ecology* 70(3): 915-924.

Lindeman, J.C & S.A. Mori. 1989. The Guianas. *In:* Campbell, D.G. & H.D. Hammond (eds.). *Floristic inventory of tropical countries.* New York Botanical Garden, New York, USA.

Longino, J.T. 1986. Ants provide substrate for epiphytes. *Selbyana* 9: 100-103.

Longman, K.A. & J. Jénik. 1987. *Tropical forest and its environment.* Longman, Essex, UK.

Lowman, M.D. & N.M. Nadkarni. 1995. *Forest canopies.* Academic Press, San Diego, USA.

Lugo, A.E. 1992. Comparison of tropical tree plantations with secondary forests of similar age. *Ecol. Monogr.* 62: 1-42.

Maas, P.J.M. & L.Y.Th. Westra. 1993. *Neotropical plant families. A concise guide to the families of vascular plants in the Neotropics.* Koeltz Scientific Books, Germany/USA.

MacArthur, R.H. & E.O. Wilson. 1967. An equilibrium theory of insular zoogeography. *Evolution* 17: 373-387.

Magurran, A.E. 1988. *Ecological diversity and its measurement.* Croom Helm, London,

Malcolm, J.R. 1991. *The small mammals of Amazonian forest fragments.* Ph.D. thesis, University of Florida, Gainesville.

Mason, D. 1996. Responses of Venezuelan understorey birds to selective logging, enrichment strips, and vine cutting. *Biotropica* 28(3): 296-309.

Mennega, E.A., W.C.M. Tammens-de Rooij & M.J. Jansen-Jacobs (eds.). 1988. *Check-list of woody plants of Guyana: based on D.B. Fanshawe's 'Check-list of the indegeneous woody plants of British Guiana'.* Tropenbos Technical Series 2. Veenman B.V., Wageningen, The Netherlands.

Migenis, L.E. & J.D. Ackerman. 1993. Orchid-phorophyte relationships in a watershed in Puerto Rico. *J. Trop. Ecol.* 9: 231-240.

Milliken, W. & J.A. Ratter. 1989. *The vegetation of the Ilha de Maraca.* Royal Botanic Garden, Edinburgh, UK.

Montgomery, G.G. & M.E. Sunquist. 1978. Habitat selection and use by two-toed and three-toed sloths. *In:* Montgomery, G.G. (Ed.). *The ecology of arboreal folivores.* pp 329-359. Smithsonian Institution Press, Washington DC.

Morellato, P.C. & H.F. Leitão-Filho. 1996. Reproductive phenology of climbers in a southeastern Brazilian forest. *Biotropica* 28(2): 180-191.

Mori, S.A. & B.M. Boom. 1987. The forest. *In:* Mori, S.A. and collaborators. The Lecythidaceae of a lowland neotropical forest: La Fumeé Mountain, French Guiana. *Mem. New York Bot. Gard.* 44: 9-29.

Mori, S.A. & G.T. Prance. 1987. Phytogeography. *In:* Mori, S.A., and collaborators. The Lecythidaceae of a lowland neotropical forest: La Fumeé Mountain, French Guiana. *Mem. New York Bot. Gard.* 44: 55-71.

Mori, S.A. 1991. The Guayana Lowland Floristic Province. *Compt. Rend. Sommaire Séances Soc. Biogéogr.* 67 (2): 67-75.

Mori, S.A., G. Cremers, C. Gracie, J.J. de Granville, M.Hoff & D. Mitchell. 1997. *Guide to the vascular plants of Central French Guiana. Part 1.* Pteridophytes, Gymnosperms, and Monocotyledons. Mem. New York Bot. Gard. 76(1): 1-422.

Myers, N. 1988. Threathened biotas: 'hot spots' in tropical forests. *The Environmentalist* 8: 187-208.

Myers, N. 1989. *Deforestation rates in tropical counties and their climate implications.* Friends of the Earth. London, UK.

Nadkarni, N.M. 1988. Tropical rainforest ecology from a canopy perspective. pp. 189-208. *In:* Almeda, F. & C.M. Pringle (eds.). *Tropical rainforests: Diversity and Conservation.* California Academy of Sciences. San Francisco, California, USA.

Naeem, S., L.J. Thompson, S.P. Lawler, J.H. Lawton & R.M. Woodfin. 1994. Declining biodiversity can alter the performance of ecosystems. *Nature* 368: 734-737.

Naeem, S., L.J. Thompson, S.P. Lawler, J.H. Lawton & R.M. Woodfin. 1995. Emperical evidence that declining biodiversity may alter the performance of terrestrial ecosystems. *Philos. Trans.* 347: 249-262.

Nelson, B.W., C.A. Ferreira, M.F. da Silva & M.L. Kawasaki. 1990. Refugia, endemism centers and collecting density in Brazilian Amazonia. *Nature* 345: 714-716.

Newbery, D.McC. 1991. Floristic variation within kerangas (heath) forest: re-evaluation of Sarawak and Brunei. *Vegetatio* 96: 43-86.

Newbery, D.McC., E.J.F. Campbell, J. Proctor & M.J. Still. 1996. Primary lowland dipterocarp forest at Danum Valley, Sabah, Malaysia. Species composition and patterns in the understorey. *Vegetatio* 122: 193-220.

Ogden, J. 1966. *Ordination studies on a small area of tropical rain forest.* M.Sc. Thesis, University of Wales, UK.

Opler, P.A., H.G. Baker & G.W. Farnkie. 1977. Recovery of tropical lowland forest ecosystems. *In:* Cairns, J., K.L. Dickson & E.E. Herricks (eds.). *Recovery and restoration of damaged ecosystems.* Univ. Press of Virginia. Charlottesville, Virginia.

Organs, G.H. 1982. The influence of tree-falls in tropical forests on tree species richness. *Trop. Ecol.* 23: 255-277.

Orians, G.H., R. Dirzo & J.H. Cushman. 1996. *Biodiversity and ecosystem processes in tropical forests.* Ecological Studies 122: 1-8.

Parker, G.G. 1995. Structure and microclimate of forest canopies. pp. 73-106. *In:* Lowman, M.D. & N.M. Nadkarni (eds.). *Forest canopies.* Academic Press, San Diego, USA.

Pellegrin, F. 1948. *Les Légumineuses du Gabon.* Librairie Larousse, Paris.

Peñalosa, J. 1984. Shoot dynamics and vegetative spread in two tropical rain forest lianas. *Biotropica* 16(1): 1-9.

Peres, C.A. 1990. Effects of hunting on on western Amazonian primate communities. *Biol. Conserv.* 54: 47-59.

Perry, D.R. 1981. The tropical rainforest canopy: a method providing total access. *Biotropica* 13: 283-285.

Phillips, O.L. & A.H. Gentry. 1994. Increasing turnover through time in tropical forests. *Science* (263): 954-958.

Phillips, O.L., P. Hall, A.H. Gentry, R.Vásquez & S. Sawyer. 1994. Dynamics and species richness of tropical rain forests. *Proc. Natl. Acad. USA* 91: 2805-2809.

Phillips, O.L. 1997. The changing ecology of tropical forests. *Biodiv. Conserv.* 6(2): 291-311.

Pinard, M., B. Howlett & D. Davidson. 1996. Site conditions limit pioneer tree recruitment after logging of Dipterocarp forests in Sabah, Malaysia. *Biotropica* 28: 2-12.

Pires, 1966 (from Campbell 1986)

Pitman, W.C., S. Cande, J. LaBrecque & J. Pindell. 1993. Fragmentation of Gwondana: the separation of Africa from South America. *In:* Goldblatt, P. (ed.). *Biological relationships between Africa and South America.* Yale University Press, New Haven & London

Pittendrigh, C.S. 1948. The bromeliad - *Anopheles* - malaria complex in Trinidad. I. The bromeliad flora. *Evolution* (Lawrence, Kansas USA) 2: 58-89.

Polak, A.M. 1992. *Major timber trees of Guyana: a field guide.* Tropenbos Series 2. Tropenbos Foundation, Wageningen, the Netherlands.

Polhill, R.M. & P.H. Raven. 1981. Swartzieae. *In: Advances in Legume Systematics Part 1.* Royal Botanic Gardens, Kew, UK.

Poncy, O., B. Riera, D. Larpin, P. Belbenoit & M. Hoff. 1996. *Floristic checklist of the Nouragues area (Spermatophytes).* Unpublished data.

Phillips, O.L., P. Hall, A.H. Gentry, R. Vásquez & S. Sawyer. 1994. Dynamics and species richness of tropical rain forests. *Proc. Natl. Acad. USA.* 91: 2805-2809.

Popma, J. & F. Bongers. 1988. The effects of canopy gaps on growth and morphology of seedlings of rain forest species. *Oecologia* 75: 625-632.

Popma, J., F. Bongers, M. Martínez-Ramos & E. Veneklaas. 1988. Pioneer distribution in treefall gaps in Neotropical rain forest; a gap definition and its consequences. *J. Trop. Ecol.* 4(1): 77-88.

Poulsen, A.D. 1996. Species richness and density of ground herbs within a plot of lowland rainforest in north-west Borneo. *J. Trop. Ecol.* 12: 177-190.

Prance, G.T. 1982. Forest refuges: Evidence from woody angiosperms. *In:* Prance, G.T. (ed). *Biological Diversification in the Tropics.* Columbia University Press, New York.

Prance, G.T. 1985. The changing forests. *In:* Prance, G.T. & T.E. Lovejoy (eds.). *Amazonia.* Pergamon Press, Oxford: 146-165.

Prance, G.T. 1989. American tropical forests. *In:* Lieth, H. & M.J.A. Werger (eds.). *Tropical rain forest ecosystems. Biogeography and Ecology.* Ecosystems of the World 14A, pp. 513-533. Elsevier, Amsterdam, The Netherlands.

Prance, G.T. 1990. The floristic composition of the forests of Central Amazonian Brazil. *In:* Gentry, A.H. (ed.). *Four Neotropical Rain Forests:* Pp. 112-140. Yale University Press, New Haven & London.

Prance, G.T., W.A. Rodrigues & M.F. da Silva. 1976. Inventario florestal de um hectare de mata de teraa firme km 30 de estrada Manaus - Itacoatiara. *Acta Amazonica* 6: 9-35.

Prance, G.T. & Kallunki. 1984. *Ethnobotany in the Neotropics.* New York Bot. Garden. New York.

Primack, R.B. 1993. *Essentials of conservation biology.* Sinaure, Sunderland, Massachusetts.

Putz, F.E. 1984a. How trees avoid and shed lianas. *Biotropica,* 16(1): 19-23.

Putz, F.E. 1984b. The natural history of lianas on Barro Colorado Island, Panama. *Ecology* 65: 1713-1724.

Putz, F.E., H.S. Lee & R. Goh. 1984. Effects of post-felling silvicultural treatments on woody vines in Sarawak. *Malaysian forester* 47: 214-226.

Putz, F.E. 1991. Silvicultural effects of lianas. *In:* Putz, F.E. & H.A. Mooney (eds.). *The biology of vines.* Cambridge University Press, Cambridge, UK.

Raaimakers, D. 1994. *Growth of tropical rain forest trees as dependent on phosphorous supply.* Tropenbos Series 11. The Tropenbos Foundation, Wageningen, The Netherlands.

Raven, P.H. & D.I. Axelrod. 1974. Angiosperm biodiversity and past continental movements. *Science* 220: 67-69.

Reid, W.V. 1994. Setting objectives for conservation evaluation. *In:* Forey, P.L., C.J. Humphries & R.I. Vane-Wright (eds.). *Systematics and conservation evaluation* 1-13. Oxford University Press.

Reitsma, J.M. 1988. *Végétation forestière du Gabon.* Tropenbos Technical Series 1. The Tropenbos Foundation, Wageningen, The Netherlands.

Richards, P.W. 1996. *The tropical rain forest. An ecological study.* Second Edition. Cambridge University Press, Cambridge, UK.

Rietkerk, M., P. Ketner & J.J.F.E. de Wilde. 1995. *Caesalpiniodeae* and the study of forest refuges in Gabon: Preliminary results. *Adansonia* 17 (1-2): 95-105.

Rodrigues, W.A. (1963). Estudo de 2,6 hectares de mata de terra firme da Serro do Navio. Território de Amapá. *Bol. Mus. Paraense Emílio Goeldi N.S.* 19 (Bot.), Belém, Brazil.

Salo, J., R. Kalliola, I. Häkkinen, Y. Mäkinen, P. Niemelä, M. Puhakka & P.D. Coley. 1986. River dynamics and the diversity of Amazon lowland forest. *Nature* 322: 254-258.

Saldarriaga, J.G. & D.C. West. 1986. Holocene fires in the north Amazon Basin. *Quat. Res.* 26: 358-366.

Sanders, H.L. 1969. Marine benthic diversity: a comparative study. *Amer. Naturalist* 102: 243-282.

Sandford, W.W. 1969. The distribution of epiphytic orchids in Nigeria in relation to each other and to geographic location and climate, type of vegetation and tree species. *Biological Journal of the Linnean Society* 1: 247-285.

Sandford, R.L., J.G. Saldariagga, K.E. Clark, C. Uhl & R. Herrera. 1985. Amazon rainforest fires. *Science* 227: 53-55.

Savill, N.J., P. Rohani & P. Hogeweg. 1997. Self-reinforcing spatial patterns enslave evolution in a host parasitoid system. *J. Theor. Biol.* 7.

Scatena, F.N. 1990. Watershed scale rainfall interception on two forested watersheds in the Luquillo Mountains of Porto Rico. *J. Hydr.* 113: 89-102.

Schimper, A.F.W. 1903. *Plant geography upon a physiological basis.* Clarendon, Oxford, UK.

Schulz, J.P. 1960. Ecological studies on rain forest in Northern Suriname. *Meded. Bot. Mus. Herb. Rijks Univ. Utrecht* 163: 1-267.

Shmida, A. & S. Ellner. 1984. Coexistence of plant species with similar niches. *Vegetatio* 58: 29-55.

Sizer, N. 1996. *Profit without plunder: Reaping revenue from Guyana's tropical forests without destroying them.* World Resources Institute. Washington DC. Via Internet: http://www.wri.org/wri/biodiv/guyana/.

Smith, T.M. & M.A. Huston. 1989. A theory of the spatial and temporal dynamics of plant communities. *Vegetatio* 83: 49-69.

StatSoft. 1993. *STATISTICA for Windows.* Release 4.5. StatSoft Inc.

Sterck, F.J. 1997. *Trees and light, tree development and morphology in relation to light availability in a tropical rain forest in French Guiana.* Ph.D. Thesis, Agricultural University, Wageningen.

Terborgh, J. 1992. *Diversity and the tropical rain forest.* Scientific American Library, New York.

Ter Steege, H. & J.H.C. Cornelissen. 1988. Collecting and studying bryophytes in the canopy of standing rain forest trees. *In:* Glime, J.M. (ed). *Methods in bryology.* Proc. Bryol. Meth. Workshop, Mainz. Pp. 285-290. The Hattori Botanical Laboratory, Nichinan. USA.

Ter Steege, H. & J.H.C. Cornelissen. 1989. Distribution and ecology of vascular epiphytes in lowland rain forest of Guyana. *Biotropica* 21 (4): 331-339.

Ter Steege, H. & C.A. Persaud. 1991. The phenology of Guyanese timber species: a compilation of a century of observations. *Vegetatio* 95: 177-198.

Ter Steege, H., V.G. Jetten, M. Polak & M.J.A. Werger. 1993. Tropical rain forest types and soils of a watershed in Guyana. *J. Veg. Sci.* 4: 705-716.

Ter Steege, H., R.G.A. Boot, L.C. Brouwer, J.C. Caesar, R.C. Ek, D.S. Hammond, P.P. Haripersaud, P. Van der Hout, V.G. Jetten, A. Van Kekem, M. Kellman, Z. Kahn, A.M. Polak, T.L. Pons, J. Pulles, D. Raaimakers, S.A. Rose, J.J. Van der Sanden, & R.J. Zagt. 1996. *Ecology and logging in a tropical rain forest in Guyana.* Tropenbos Series 14. The Tropenbos Foundation, Wageningen.

Ter Steege, H. 1993. *Hemiphot, A programme to analyze vegetation indices, light and light quality from hemispherical photographs.* Tropenbos Documents 3. Tropenbos Foundations, Wageningen.

Thiollay, J.M. 1989. Area requirements for the conservation of rain forest raptors and game birds in French Guiana. *Conserv. Biol.* 5: 18-32.

Tilman, D. & J.A. Downing. 1994. Biodiversity and stability in grasslands. *Nature* 367: 363-365.

Tobin, J.E. 1995. Ecology and diversity of tropical canopy ants. *In:* Lowman, E.D. & N.M. Nadkarni (eds.). *Forest canopies.* Academic Press, San Diego, pp. 129-147.

Tolmatschew, A.I. 1971. Uber einige quantitative Wechselbeziehungen der Floren der Erde. *Feddes Repert.* 82 (5): 343-356.

Tropenbos 1996. *Towards priorities of biodiversity research in support of policy and management of tropical rain forests. A contribution to the conservation and wise use of tropical forests.* The Tropenbos Foundation, Wageningen.

Tropenbos 1997. Internet site. *http://www.tropenbos.nl*

Tuomisto, H. & K. Ruokolainen. 1997. The role of ecological knowledge in explaining biogeography and biodiversity in Amazonia. *Biodiv. Conserv.* 6: 347-357.

Uhl, C., K. Clark, N. Dezzeo & P. Maquirino. 1988. Vegetation dynamics in Amazonian treefall gaps. *Ecology* 69: 751-763.

Valencia, R., H. Balslev & G.C. Paz y Mino. 1994. High tree alpha-diversity in Amazonian Ecuador. *Biodiv. Cons.* 3: 21-28.

Van der Hout, P. 1996. *Effects of logging with different intensities of low impact harvesting.* Tropenbos-Guyana Reports 96-1. Tropenbos-Guyana Programme, Georgetown, Guyana.

Van der Laan, J.D. & P. Hogeweg. 1995. Predator-prey coevolution: interactions across different time scales. *Proc. Royal Soc. London B* 259: 35-42.

Van der Meer, P.J. & F. Bongers. 1996a. Patterns of treefalls and branchfalls in a tropical rain forest in French Guiana. *J. Ecol.* 84(1): 19-30.

Van der Meer, P.J. & F. Bongers. 1996b. Formation and closure of canopy gaps in the rain forest at Nouragues, French Guiana. *Vegetatio* 126(2): 167-180.

Van Deursen, W.P.A. & C.G. Wesseling. 1993. *The PC-Raster package.* Department of Physical Geography, University of Utrecht.

Van Kekem, A.J., J.H.M. Pulles & Z. Kahn. 1996. *Soils of the rainforest in Central Guyana.* Tropenbos - Guyana Series 2. pp. 160.

Van Schaik, C.P. & E. Mirmanto. 1985. Spatial variation in the structure and litterfall of a Sumatran rainforest. *Biotropica* 17: 196-205.

Vásquez, R. & A.H. Gentry. 1989. Use and misuse of forest-harvested fruits in the Iquitos area. *Conserv. Biol.* 3: 350-361.

Veneklaas, E.J. 1990. Nutrient fluxes in bulk precipitation and throughfall in two montane tropical rain forests, Colombia. *J. Ecol.* 78: 974-992.

Veneklaas, E.J. & R. Van Ek. 1990. Rainfall interception in two tropical montane rain forests, Colombia. *Hydr. Proc.* 4: 311-326.

Veneklaas, E.J. 1991. Litterfall and nutrient fluxes in two montane tropical rain forests, Colombia. *J. Trop. Ecol.* 7: 319-336.

Veneklaas, E.J., R. Zagt, A. Van Leerdam, R. Van Ek, G. Broekhoven & M. Van Genderen. 1990. Hydrological properties of the epiphytic mass of a montane tropical rain forest, Colombia. *Vegetatio* 89: 183-192.

Vitousek, P.M. & J.S. Denslow. 1986. Nitrogen and phosphorus availability in treefall gaps of a lowland tropical rain forest. *J. Ecol.* 74: 1167-1178.

Vitousek, P.M. & D.U. Hooper. 1993. Biological diversity and terrestrial ecosystem biogeochemistry. *In:* E.D. Schulze & H.A. Mooney (eds.) *Biodiversity and ecosystem function.* Pp. 3-14. Springer Verlag, Berlin.

Voorhoeve, A.G. 1964. Some notes on the tropical rainforest of the Yoma-Gola National forest near Bomi Hills, Liberia. *Commenwealth Forestry Review* 43: 17-24.

Waide, R.B. & A.E. Lugo. 1992. A research perspective on disturbance and recovery of a tropical montane forest. *In:* Goldammer, J.G. (ed.). *Tropical forests in transition.* Birkhäuser, Basel.

Ward, H.H. 1963. Hierarchical grouping to optimise an objective function. *J. Amer. Stat. Assoc.* 58: 236-244.

Waterman, P.G., C.N. Mbi, D. McKey & J.S. Gartlan. 1980. African rain forest vegetation and rumen microbes: phenolic compounds as correlates of digestibility. *Oecologia* 47: 22-33.

Waterman, P.G. & D. McKey. 1990. Herbivory and secondary compounds in rain-forest plants. *In:* Lieth, H. & M.J.A. Werger (eds.). *Tropical rain forest ecosystems. Biogeography and Ecology.* Ecosystems of the World 14B, pp. 513-533. Elsevier, Amsterdam, The Netherlands.

Welch, I.A. & G.S. Bell. 1971. Great Falls Inventory. *Forest Department Report* no. 41/70. Forest Department, Georgetown, Guyana.

Welden, C.W., S.W. Hewett, S.P. Hubbell & R.B. Foster. 1991. Sapling survival, growth and seedling establishment: relationship to canopy height in a neotropical forest. *Ecology* 72: 35-50.

Went, F.W. 1940. Soziologie eines tropischen Urwaldes. *Ann. Jard. Bot. Buitenzorg* 50: 1-98.

Werger, M.J.A. 1973. Notes on the phytogeographical affinities of the southern Kalahari. *Bothalia* 11: 177-180.

Werger, M.J.A. 1992. De aantasting van het regenwoud. *In:* Hoogmoed, M.S. & R. de Jong (eds.). *Tropisch regenwoud: schatkamer van biodiversiteit.* Nationaal Natuurhistorisch Museum, Leiden.

Williams, P.H. & K.J. Gaston. 1994. Measure more of biodiversity: can higher-taxon richness predict wholesale species richness? *Biol. Conserv.* 67: 211-217.

Williams, P.H. & C.J. Humphries. 1996. Comparing character diversity among biotas. *In:* Gaston, K.J. (ed.). *Biodiversity. A biology of numbers and differences.* Blackwell Science, Oxford.

Wilbur, R.L. 1994. Vascular plant: an interim checklist. *In:* McDade, L.A., S.B. Kamaljit, H.A. Hespenheide & G.S. Hartshorn (eds.). *La Selva: ecology and natural history of a Neotropical rainforest.* University of Chicago Press, Chicago, USA.

Wilson, E.O. 1992. *The diversity of life.* Belknap, Cambridge, UK.

Whitmore, T.C. 1984. *Tropical rain forests of the Far East.* Clarendon Press, Oxford.

Whitmore, T.C. 1989. Canopy gaps and the two major groups of forest trees. *Ecology* 70(3): 536-538.

Whittaker, R.H. 1967. Gradient analysis of vegetation. *Biol. Rev.* 42: 207-264.

Whittaker, R.H. 1977. Evolution of species diversity in land communities. *Evol. Biol.* 10: 1-67.

Wolf, J.H.D. 1994. Factors controlling the distribution of vascular and non-vascular epiphytes in the northern Andes. *Vegetatio* 112: 15-28.

Wright, S.J. 1992. Seasonal drought, soil fertility and the species density of tropical forest plants communities. *TREE* 7: 260-263.

Wright, S.J. 1996. Plant species diversity and ecosystem functioning. *In:* Orians, G.H., R. Dirzo & J.H. Cushman (eds.). *Biodiversity and ecosystem processes in tropical forests.* Ecological Studies 122. Pp. 11-31. Springer Verlag. Berlin.

Zagt, R.J. 1997. *Tree demography in the Tropical Rain forest of Guyana.* Ph.D.-thesis, University of Utrecht. Tropenbos-Guyana Series 3. Elinkwijk, Utrecht.

Zagt, R.J. & M.J.A. Werger. In press. Community structure and the demography of seedlings and adults in tropical rain forests, with examples from the Neotropics. *In:* Newbury, D.M., H.H.T. Prins & N.D. Brown (eds.). *Population and community dynamics in the tropics.* British Ecological Society Symposium Volume No. 37. Blackwell Scientific Publishers, Cambridge.

Acknowledgements

Hoe het toch allemaal zo ver kon komen......

De tropen..........mijn lokroep om naar de tropen te gaan is waarschijnlijk, net als bij anderen, voortgekomen vanuit thuis. De mijmerende verhalen van mijn vader: de baboe, kubangspatoe, klapper, je 'senang' voelen..... allemaal exotische klanken, uitgesproken met een ondertoon van heimwee. Foto's van prachtige landschappen, statige dames en heren op een open veranda, m'n grootouders zo duidelijk niet meer 'Hollands'.

Frans Guiana blijft m'n eerste tropische liefde, en specifiek het bos rond Saül, een echt Amazonisch gemengd regenwoud. Maar ja, wat kan er nu op tegen een uitzicht vanuit een emergente boom die na twee dagen geduldig proberen eindelijk bedwongen is met behulp van pijl en boog, klimtouwen en klimuitrusting? Guyana was in eerste instantie heel anders, een kleinschalig mozaïek van bostypen gecorreleerd met het voorkomen van, veelal erg arme, bodems. Mijn eerste ervaring in het Mabura Hill gebied was samen met Marion Jansen-Jacobs, Eimear NicLughadha en Hans ter Steege. Hans joeg ons in een noodtempo door het reservaat, en we mochten geeneens verzamelen! Dat heeft m'n beeld vrij lang bepaald. In 1991 kwam ik terug om samen met Hans (en Koos) een artikel over epifyten te schrijven, iets wat pas in 1997 is afgerond (hoofdstuk 3, dit proefschrift). Toen heb ik wat meer van het bos gezien, zag verschillende Tropenbossers aan het werk, en werd zeer gefascineerd door de verscheidenheid aan vegetaties. Toen ik eenmaal D.B. Fanshawe met veel liefde had horen vertellen over 'zijn' Guyana was mijn interesse helemaal gewekt.

Maar genoeg over de voorgeschiedenis, dit gaat over dit proefschrift en er zijn velen die een bijdrage hebben geleverd aan de totstandkoming van dit boekje.

Bedankt!

Concerning my stay in Guyana there are many people I wish to thank. I would like to start with Sam Roberts. It was an enjoyment working with him and to be able to share his knowledge of the species of the forest and 'bush life'. His contributions to this study are manyfold, including his suggestion to employ his son, George, as my first field assistant. Thanks are due to all those that worked with me in the field and/or enlivened my stay: George, Maurice, Martin, Souky, Henry, the Pibiri crew (s), and in particular Bells, Julian, Monty, Chris, Isaac, Charlton, Dane, Ryan, Pilay and Hoppy.

All Tropenbos colleagues, students and counterparts are thanked for their contribution in, and sharing of, laughter and sorrows, beer and pringles, and bubbling and w(h)ines: Allison, Arjan, Andre, Barbara, Bettine, Carmen, Coralie, Cornel, David, Dorinne, Ferry, Hans, Joost, Julian, Juul, Leo, Lonneke, Karl, Kristel, Liesbeth, Marissa, Mathilde, Menno, Michelle, Niels, Nienke, Ramesh, Raquel, René, Roderick, Paddy, Peter, Petra, Simmone, Tinde, Victor, and Zab.

Special thanks are also in place for Elaine and Donna, whose friendship, neck-massages, prayers and 'home-remedies' kept me going in hard times. I'll always keep fond memories of Elaine as 'campie'. Alexis, my god child, will remain a permanent bond with Guyana.

In Georgetown, that famous place with a telephone, there were always Sybille and Joan to brighten your day, and first Harry, later Maina and Colette, who carried the burden of running the logistics of the Tropenbos-team. Kissoon was the go-between.

En dan komt de verwerkingsfase. Om te beginnen zijn daar mijn promotoren, Marinus Werger en Paulien Hogeweg, zonder wie dit boekje er vast heel anders had uitgezien. Twee promotoren: het betekende in de praktijk één voor de theorie en het snelle meedenk vermogen, en één voor de opzet, de 'finishing touch' en het leren (?) helder formuleren. Ik wil speciaal Paulien bedanken voor haar enorme 'mind boggling' vermogen, ik stond keer na keer verbaasd over haar vermogen om over te schakelen van de ene biologische discipline naar de andere. Elke keer wist ze dan nog precies wist waar we de laatste keer met onze discussie gebleven waren. Zonder haar belangstelling en stimulerende discussies had dit boekje er misschien niet gelegen, en ik heb goede herinneringen aan haar bezoek aan Mabura. Marinus is, gedeeltelijk samen met anderen, als een ware drakendoder ten strijde getrokken tegen mijn creaturen van zinsconstructies en 'gallopeer-nijgingen', dus als u nog steeds vindt dat dit boekje niet leesbaar is, aan hem heeft het niet gelegen.

Vervolgens een speciaal woord van dank voor mijn co-auteurs: Hans ter Steege, Koos Biesmeijer, Juul van Dam, Cornel van der Kooij en Peter van der Hout, en voor Andre Noest die de algoritmen voor het berekenen van een verhoogd lichtniveau n.a.v. de positie en grootte van gaps heeft gemaakt. Ik wil ook graag diegenen bedanken die, naast mijn beide promotoren, naar de verschillende manuscripten gekeken hebben: Paul Maas, Hiltje Maas, Roderick Zagt, Padmatti Harripersaud, Hans ter Steege, Marjet van Houten, Erik Mennega en Victor Jetten.

Paul Maas, de hoogleraar van de groep die mijn 'thuisbasis' vormde, is de volgende die ik graag wil bedanken. Zijn grote kennis betreffende de herkenning van Neotropische soorten, gepaard met een enthousiaste Neotropische 'zendingsdrang', waren een belangrijke stimulans tijdens mijn studie en de voorbereiding van dit onderzoek. Zijn inbreng op de meer floristische kant van dit onderzoek, zijn warme, persoonlijke belangstelling én zijn bezoek aan Mabura zijn ook zeer wezenlijk geweest. Ik vind het dan ook onterecht dat je niet onder drie professoren kunt promoveren.

Marion Jansen-Jacobs, die de Nederlandse kant van de besognes met 'mijn' studenten voor haar rekening nam, was ook op meerdere punten belangrijk voor dit onderzoek. Haar bezoek aan de praktijk van het opnames maken vergrootte de prettige samenwerking rond de begeleiding van de studenten. Daarnaast zal haar essentiële onderscheid tussen ballen- en balletjes bomen, zoals ook tussen knoedeltjes, frutseltjes, en frumseltjes altijd een rol blijven spelen in mijn familie-herkenning (samen met dergelijk gebromde kenmerken van Jan Lindeman, natuurlijk). Marion was m'n contactpunt met de thuisbasis ten tijde dat ik in Guyana zat en de informatie die ik via haar (en Ara Görts-van Rijn en Hiltje Maas) ontving, gaven me het gevoel dat ik er nog bij hoorde.

En dan zijn er natuurlijk 'mijn' studenten. Om te beginnen had ik zonder hen nooit die ontzettend grote data-set kunnen verwerven. Daarnaast hebben ze me geholpen met het opzetten van het opname-systeem en het bedenken van de code-namen (Barbara, Karl & Lonneke), om snel een eerste indruk te krijgen van de verschillen tussen natuurlijk en geëxploiteerd bos (Arjan & Ferry) en met het meer in detail bestuderen van de lianen en met het toepassen van GIS op kleinschalig niveau (Juul & Cornel, Niels & Menno). Maar ze waren voor mij ook de 'krenten-in-de-pap'. Het was keer op keer leuk om over met studenten het bos in te gaan voor wie het tropisch bos en haar verrassingen nieuw waren, waardoor ik ook weer even met een andere bril naar het bos keek en weer al m'n 'juffigheden' kwijt kon. Ik hoop dat ik ze niet al te veel heb doorgezaagd over waaraan je ook al weer een Julian Rope of Climbing Baromalli kan herkennen. Ik heb in ieder geval veel plezier beleefd aan het gezamenlijke veldwerk en de discussies tijdens veld- en uitwerkperiodes. Wim Dijkman, Heinjo During en Steven de Jong hebben door hun ondersteuning aan de studenten bijgedragen aan de tot standkoming van dit proefschrift.

Een speciaal dankwoord mag ook niet ontbreken voor Bep Mennega. Ze was er altijd op de cruciale momenten om over mij haar (financiële) steun te verlenen, via haar 'Alberta Mennega Fonds', of persoonlijk. De door mij gevolgde GIS-cursus, nodig voor het goed uitwerken van de gegevens, had ik zonder haar dan ook niet kunnen volgen.

Ben ter Welle played an important role initiating this research and in the beginning he was the Dutch coordinator of the Tropenbos-Guyana Programme. A task later performed by Thijs Pons. George Walcott, Indarjit Ramdass, René Boot and Hans ter Steege (and in between also shortly Leo Brouwer) dealt with the Guyanese side of Tropenbos business. Without them this research would not have been possible. Thanks to all of them.

Rob Bakker, Gerrit Cornelissen, Erik Simonis, Huub Favier en Esmeralda van Rhenen zorgden gezamenlijk voor de essentialia rond het planten verzamelen, drogen, ontsmetten, opslaan, administreren en versturen. Guda Lichthart verrichtte wonderen met het aanvragen van literatuur. Hartelijke dank voor alle hulp.

Werkend bij drie vakgroepen zijn er nog zoveel mensen meer die hier een plekje verdienen: Jifke, Els, Bep, Lubbert, Lars, Paul, Hiltje, Ben, Doortje, Marion, Jan, Hendrik, Arian, Tinde, Gea, Ara, Erik, Ad, Guda, Esmeralda, Erik, Rob & Huub zorgden voor heel wat aangename uren bij het herbarium. Verder wil ik iedereen bedanken bij de vakgroep Theoretische Biologie en afdeling Vegetatie Oecologie.

Marjolein Kortbeek-Smithuis (opmaak) en Bart Landman (omslag) zorgden voor het eindresultaat in de vorm van dit boekje. De serie wordt steeds mooier!

En dan was er natuurlijk het thuisfront: in de eerste plaats Annemiek en Marjet. Meiden, bedankt voor alles! Ik kon jullie altijd bellen, jullie lieten je niet van de wijs brengen door de echo op de lijn en ik geloof dat mijn Georgetownse telefoonrekeningrecord nog niet is verbroken. René Wessels was mijn financiële zaakwaarnemer, waardoor ik met een gerust hart in het buitenland kon zijn. Daarnaast zijn de trouwe brieven van Edith, AnneMarie, Marieke & René en Marianne een permanente 'life-line' geweest. Het bezoek van AnneMiek, Anneke, Marianne en Nienke, kerst 1995, gaf me de kans om over nog één maal langs de inmiddels vertrouwde plekjes te gaan en mijn ervaringen te delen. Bij terugkomst bleek hoe essentieel zoiets is.

Pa & Ma, jullie hebben me het bewijs gegeven dat jullie me door dik en dun steunen (en nooit meer verbaasd zullen staan over waar ik allemaal mee aan kom). Ik ben trots op jullie!

Pip, plak, tif, dia en scan: allemaal kleine woorden die bij een groot(s) iemand horen. In ieder geval te groots voor een goed dankwoord op deze plek.

Curriculum Vitae

Renske C. Ek werd geboren op 24 december 1960 te Dordrecht. De middelbare school werd in etappes doorlopen: in 1977 behaalde zij het MAVO diploma aan de Prinses Beatrix Mavo te Lekkerkerk, in 1980 het HAVO diploma aan de Gemeentelijke Pedagogische Academie te Rotterdam en in 1982 behaalde zij het diploma Atheneum-B aan de scholengemeenschap 'De Krimpenerwaard' te Krimpen aan de IJssel.

In datzelfde jaar startte zij met de studie Biologie aan de Universiteit Utrecht. Het doctoraalexamen werd in januari 1989 behaald. Tijdens de studie werden twee afstudeervakken gedaan. Het eerste onderwerp betrof de verticale verdeling van epifyten over bomen in het tropisch regenwoud van Frans Guiana. Dit afstudeervak vond gedeeltelijk plaats bij het Instituut voor Systematische Plantkunde (later afdeling Herbarium) en bij de afdeling Botanische Oecologie. Het tweede onderwerp betrof verdere patroonherkenning van de epifytische dataset bij de vakgroep Theoretische Biologie.

Tussen 1989 en 1992 was zij op basis van een tweetal korte contracten werkzaam als toegevoegd onderzoekster bij de afdeling Herbarium, vakgroep Botanische Oecologie en Evolutie Biologie, Universiteit Utrecht. In deze periode heeft ze zich verdiept in de geschiedenis van het planten verzamelen in de Guianas en heeft zij voor de 'Flora van de Guianas' twee deeltjes geschreven, een index van verzamelaars in Guyana en een index over Suriname.

Sinds augustus 1992 werkte zij in opdracht van de Stichting Tropenbos en de EU aan het onderzoek dat leidde tot dit proefschrift. Dit onderzoek werd uitgevoerd als toegevoegd onderzoekster bij de afdeling Herbarium, vakgroep Botanische Oecologie en Evolutie Biologie, Universiteit Utrecht. De data-analyses voor dit onderzoek werden uitgevoerd bij de vakgroep Theoretische Biologie.